Forest Farm Husbandry

Matthew Fedden

Practical
ACTION
PUBLISHING

TECHNOLOGY CONSULTANCY CENTRE, 1988

Practical Action Publishing Ltd
25 Albert Street, Rugby,
Warwickshire, CV21 2SD, UK
www.practicalactionpublishing.com

A catalogue record for this book is available from the British Library & Library of Congress

ISBN 10: 1 85339 006 2
ISBN 13: 9781853390067
ISBN Digital book: 9781780442211

Citation: Fedden, M. (1988) *Forest Farm Husbandry*, Rugby, UK: Practical Action Publishing https://doi.org/10.3362/9781780442211

Since 1974, Practical Action Publishing has published and disseminated books and information in support of international development work throughout the world. All print editions are produced and distributed via ethical and sustainable print on demand global facilities.

Practical Action Publishing is a trading name of Practical Action Publishing Ltd (Company Reg. No. 01159018 | VAT 880 9924 76). All profits are covenanted back to its parent group, Practical Action (Charity Reg. No. 247257).

The manufacturer's authorised representative in the EU for product safety is Lightning Source France, 1 Av. Johannes Gutenberg, 78310 Maurepas, France. compliance@lightningsource.fr

Foreword

When Dr F W Lukey, then Deputy Director of Ghana's Technology Consultancy Centre (TCC), first produced his handbook on *Minimum Tillage Farming*, the Centre's on-farm experience was limited to only one year of two farming seasons. Our knowledge was mainly derived from the literature, visits to IITA in Nigeria and Dr Lukey's five-year computer-aided study of the economics of farming options in Ashanti.

Now the situation is changed considerably. The TCC experience now embraces more than five years on-farm practical minimum tillage farming in the forests of Ashanti. Thus we can produce with greater confidence a book offered by farmers for farmers. We have decided to produce a new and entirely revised version of our *Minimum Tillage Farming* handbook and we have even given it a new name.

Minimum tillage farming at TCC has always been an Anglo-Ghanaian undertaking. Of the Englishmen, after Frank Lukey came Bob Moss and Matthew Fedden. On the Ghanaian side Kwesi Opoku-Debrah, Kwesi Kyere and Evans Dawoe all made their contributions. But it fell to Matthew Fedden to distil the experience and pass on the fruits of our labours for the benefit of others.

Matthew Fedden and Bob Moss before him were British Voluntary Service Overseas (VSO) personnel. Acknowledgement must be made to the VSO organization for this invaluable contribution to the project.

During the past two years, Matthew Fedden and Evans Dawoe and their team have introduced alley cropping on the TCC Minimum Tillage Pilot Project Farm at Fumesua. This system seems to offer great promise as a soil conserving and enriching technique that also affords good weed control. For these reasons this handbook focuses

i

more on alley cropping than the other basic minimum tillage techniques featured in Frank Lukey's earlier work. We hope that by encouraging farmers to adopt alley cropping they will not only derive benefits in terms of crop yields and cash income but also contribute towards replacing the essential tree cover, much of which has been denuded. In fact, alley cropping can be viewed as a variation of the traditional bush fallow system in farming in which the trees replenish the soil but under controlled conditions of continuous cropping. Thus the traditional forest farmer can regard the new system as a logical extension of what he has known and practised. He learns to work with the same natural forces but with a new intensity and at an accelerated pace which yields benefits that are both economic and ecological.

Some clients of TCC are already practising alley cropping. To these we wish every success and hope that many will follow their brave pioneering ventures. In preparing this book, Matthew Fedden has done much to encourage this trend.

JOHN POWELL, Kumasi
September 1986

Contents

CHAPTER 1

The Case for Minimum Tillage and Alley Cropping in the Forest Zone of Ghana

With her rapidly increasing population, one of Ghana's most pressing concerns has been, and will, for the foreseeable future, be the need to feed her people. Ghana's climate and soils have great productive potential and the need for food must be met from within. The problem is most serious in towns where access to land is limited and where the number of mouths to feed is greatest. Such a situation results in the land around towns being rapidly overused and depleted, forcing food production into remoter areas.

By far the majority of farms in the forest zone of Ghana are smallholdings farmed with cutlass, hoe and a match. Because of the rate at which plants grow in Ghana, every farmer is fighting a constant battle against weeds which will otherwise choke the crop. This battle takes up great amounts of time during the growing season and, the larger the farm, the longer it will take. In practice the traditional cutlass and hoe farmer is severely limited in the area that can be cultivated by the need to control weed growth. The production capacity of a farmer is directly related to the area that can be cultivated. The traditional farmer operates a way of agriculture ideally suited to providing food for a small community well supplied with accessible land. It is an extensive system which grows crops for three or four years and then operates a fallow for considerably longer. It fails when the community becomes a town, as the scale of production is too great for both the agricultural system and the land. The fallow period is the first to suffer, and the soil rapidly becomes worn out through overcropping. The result is vast areas of exposed laterite and sparse scrub often seen surrounding towns. A temporary solution is for the farmer to farm more distant land. The distance travelled increases demands on the farmers time during the growing season, reducing productivity whilst spoiling more land. These then are the problems to which an appropriate solution is being sought.

It was thought that mechanization offered a solution, increasing productivity at a stroke. Such a solution takes no account of the majority of Ghana's farmers: very few can afford a tractor. The resulting machinery compaction and tillage takes no account of the notoriously fragile nature of the soil, nor the numerous obstacles such as tree roots, which damage expensive machinery in the forest zone. Finally it relies on a good infrastructure to service and maintain the machinery. This is lacking in many parts of Ghana. It has some success further north in the Ghana midlands, these being more suited to mechanized tillage. These areas are a long way from the demand, and transport is expensive. The northern areas enjoy a less favourable growing climate and have lower potential yields than the forest zone.

Minimum tillage is a widely used farming technique that can be adapted to provide a solution. It is based on the principle that the land should be disturbed as little as possible, neither ploughed nor hoed. Weeds are controlled by the use of chemicals and mulching. In the forest, chemical application is best done by one man with a relatively cheap knapsack sprayer, rather than a tractor-mounted unit. Dead weeds are left on the soil to provide a protective cover or mulch. This inhibits further weed emergence, and it also protects the soil against erosion caused by heavy rainfall. Eventually, the mulch will break down and add to the soil fertility. When planting, the seed has to be sown in the untilled soil. Planting implements vary in complexity from a cutlass to a tractor-drawn pneumatic seed drill. TCC advocates the use of the Rolling Injection Planter. This simple tool is made in Kumasi and Tamale, and can readily be manufactured by workshops in other localities. It is hand-pushed by one person. The area planted is much greater than can be achieved by the use of the cutlass alone, experience showing that two acres can be covered in four hours. Obstacles such as stumps can easily be avoided by this wheelbarrow-like machine. Laying mulch down between the crop rows will prevent much weed emergence. The few weeds that do emerge are easily dealt with.

Using minimum tillage techniques, the small-scale farmers are able to control the weed problem, and thus expand their farm size. Minimum tillage is a good way of conserving soil fertility. It is designed for continual cropping, therefore land is more productive. When minimum tillage techniques are used in conjunction with beneficial (leguminous) crops such as cowpeas, it is possible to reclaim previously barren land. (This feat has been demonstrated at Approtech Farms Ltd, Kumasi, where TCC has been involved with a long running minimum tillage programme.) All these features can improve the productivity of both the small-scale farmer and the land being farmed. The techniques are ideal for farms close to towns where the demand for food is greatest. So the food is being grown right where it is needed.

There are foreign exchange costs involved with minimum tillage farming. Such things as fertilizers, chemicals and sprayers are currently all imported. Any technological advance from traditional agriculture is difficult, if not impossible to envisage without some foreign exchange expenditure on agricultural inputs. In this case the money would be invested into a system that has been proven to work in the Kumasi region, rather than a mechanized system that has been repeatedly demonstrated not to be self-supporting in the forest zone. It would benefit large numbers of small-scale farmers rather than a few well-off farmers. It would also benefit the environs of towns as well as providing their food.

In the context of the present drive to revive the cash cropping of perennial crops such as cocoa and oil palm, minimum tillage does not involve burning. Every year bush fires destroy and set back huge areas of productive cocoa and palm plantations. Minimum tillage is compatible with farming perennial crops. Unlike methods relying on burning or soil disturbance, it poses no threat to the crop, and can be used confidently within vulnerable young plantations, before the trees attain maturity.

Alley cropping is the new form of a traditional technique in Ghana: agroforestry. Currently it is receiving attention from scientists throughout the humid tropics. It is a technique of managing a farm so that land fertility is protected and enhanced while need for the imported inputs necessary to grow crops on the same land is reduced year after year. TCC has been practising minimum tillage on an alley cropping plot, on a farm near Kumasi. Selected species of trees have been planted, which fix large amounts of nitrogen in the soil. These trees are planted in rows amongst the crops, and the canopy shade during fallow periods (dry season and between crops) is an effective weed suppressor. During cropping the trees are cut back to bare poles, to minimize competition between the crop and the trees. The

3

trees provide an effective barrier to erosion throughout the year. The roots are permanently binding the soil, and this is coupled to the protective effects of either a canopy or a mulch. Nitrogen is provided to the soil, which will reduce, but not abolish, dependence on added fertilizer or manure. Shade provided by the canopy before planting the crop allows reduced use of weedkillers compared to conventional minimum tillage. The technique is particularly suitable for small farms requiring a sustainable yield without a heavy dependence on chemicals. Yields will be relatively stable, reflecting the soil protecting, weed suppressing and nutrient recycling characteristics of the system.

CHAPTER 2

Soil Fertility

Soil fertility varies enormously in the humid tropics, even within one field. Though a very few crops can grow in poor soils, it is common sense to keep agricultural land fertile, in order to enable a wide range of crops to be grown. Most of the fertility of a soil lies in the top few inches. It is vital to protect this layer.

Fig. 1. Badly eroded soil at Basumptwe.

Erosion

This is generally associated with bare soil. Land should never be burnt and left bare. The first rainstorm may result in the immediate loss of the topsoil. In the same way that a stream will dig itself a bed, so water running off a field will quickly dig down into the lower soil, washing away the fertile topsoil. Erosion problems become severe where the field slopes.

Overcropping

Every plant that grows on a field will take food from the soil. When the farmer plants high densities of crops such as maize, yams, and vegetables, these take a great deal of soil nutrient reserves in order to produce a harvest. Unless these food reserves are replaced, the soil will become gradually poorer until it is able only to sustain very low yields. Formerly this problem was solved by moving to another piece of land, leaving the denuded soil to recover, usually for a period of six or seven years. Where farmers own the land they work this rapidly becomes uneconomic. Any serious attempt at commercial food production must conserve soil fertility.

Overheating

The sun in Ghana is so powerful that it can bake the soil like a loaf of bread. If the soil surface is left bare, as after planting, the heat of the sun can kill off all but the most hardy weeds. Such temperatures will also cause the chemical breakdown of the soil's nutrient reserves.

Leaching

The topsoil contains plant food: nitrogen, phosphorus, potassium, and other nutrients, by binding them closely to soil particles. Some of these compounds, particularly the nitrates, are quickly dissolved in water. Rain water percolating down through the soil will tend to carry plant nutrients with it, out of the reach of crop roots.

Maintenance of soil fertility

Many of these problems can be solved by good management practices. The most important rule is that the soil should not be left bare. Also, burning or hoeing a soil only exposes it to the damaging effects of the sun and rain. Weeds, when killed, should be laid on the soil surface to protect it and prevent it from drying out. The same should be done with crop residues. This practice is called mulching, which is the most important feature of minimum tillage farming.

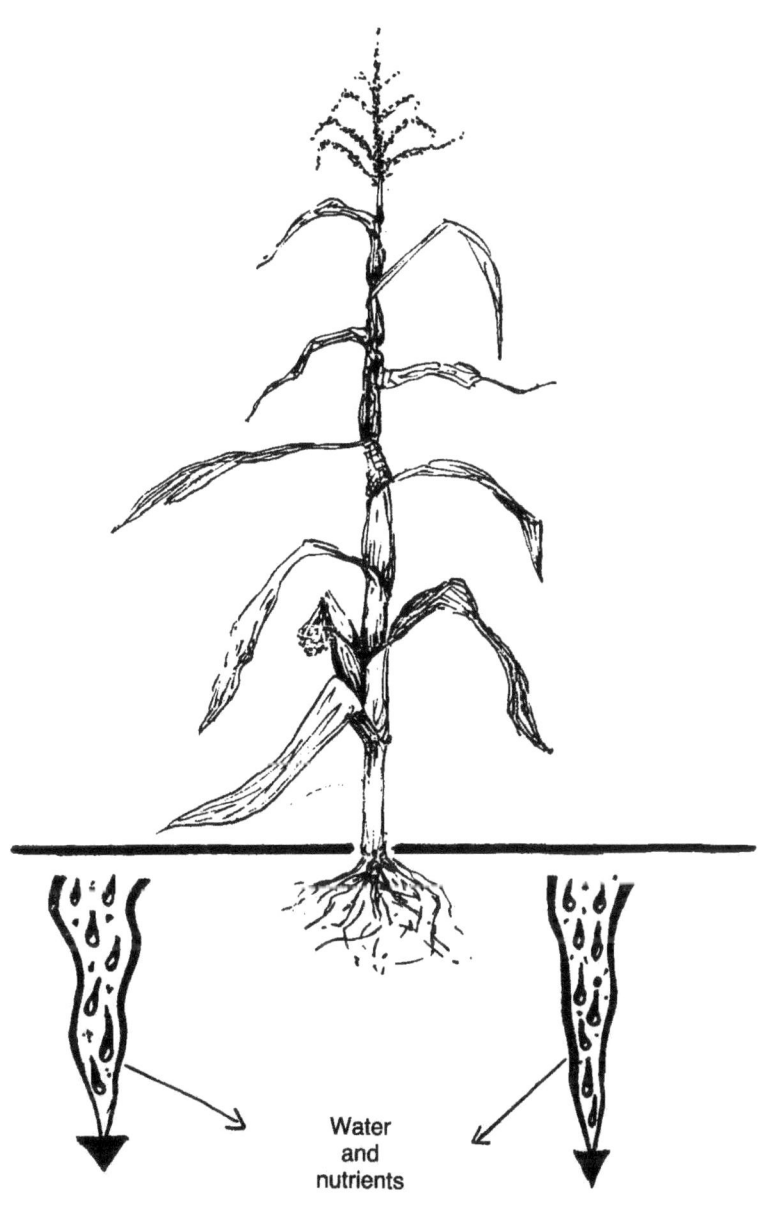

Water
and
nutrients

Fig. 2. Diagram of leaching effect.

Soil erosion, which is caused by allowing water to run off bare soil, can be minimized by leaving the roots of dead weeds in the soil. The roots hold the soil together and prevent rain water from prising loose the soil particles. When the roots rot, they leave small holes through which surface water can percolate into the subsoil. The presence of a layer of mulch on the soil surface will also help to minimize erosion. A heavy mulch will prevent the hammer-hard raindrops from eroding the topsoil, and it will slow down the progress of the surface water as it runs down the field. As the water moves more slowly over the soil surface, so more will percolate down into the soil, adding to the water available to growing crops, and less remains to prise loose the top soil.

Increasing soil fertility

A balanced rotation is needed for farmers who continuously crop the same land, to put some of the nutrients back into the soil that the demanding crops take out. Nitrogen-fixing (leguminous) plants should form a regular part of this rotation. The use of a mulch will also help, as it will decompose and enrich the soil.

Mulching

The use of a mulch will buffer the soil against sudden temperature changes. It will also preserve soil moisture during short dry spells in the cropping season.

Improving soil fertility

Generally the two nutrients most limiting crop production in Ghana are Nitrogen and Phosphorus. Every season it will be necessary to provide crops with these compounds. This can be done by chemical fertilizer, or organic manure, or a mixture of both.

Fertilizers

These chemicals supply the main food requirements of crops in concentrated form. Currently three types are commonly available in Ghana. These are: 20:20:0 which contains nitrogen and phosphorus, and ammonium sulphate which is a concentrated form of nitrogen. Urea, another highly concentrated form of nitrogen, may also be available in some areas.

The use of a mulch will encourage crop roots to grow near the soil surface. This allows the farmer to place fertilizer on top of the soil. In conventional agriculture, fertilizer must be incorporated into the soil in order to be taken up by the roots. This represents a considerable saving of time and effort.

Manure

Many farms do not make good use of the natural resources available. Instead of spreading manure onto their soils they will pay money to buy chemical fertilizers which may not be as good. A rich manure provides all that a fertilizer can and more, as it is a soil conditioner. It will add nitrogen, phosphorus, and potassium, as well as trace elements needed by the crop plants. It also contains organic matter which enables the topsoil to bind these nutrients, and enhances root growth.

Manure from any animal can be used. Chicken manure is one of the best, but sheep, goat, pig and cow manure are also good. When using manure containing a lot of bedding sawdust or straw, this should be allowed to rot down before it is applied onto the field. Cow manure also should be allowed to rot well. This will guard against the germination of noxious weed seeds of plants such as *Rottboelia exaltata* (Twi-*nkyenkyenma*) which the cattle may have eaten. Compost is similar to manure in its benefits though it takes some effort to prepare. (See Appendix 1 for details on how to make compost.)

Land reclamation

Sandy soils often yield poorly because they are unable to bind nutrients tightly. Consequently, leaching is severe. Such soils require organic matter, and will benefit most from applications of manure and least from chemical fertilizer. A long-term method of improving sandy soils is to work clay or silt into the surface. The opposite is true of clay soils where sand spread over the soil surface will gradually improve the soil quality. If the topsoil has been removed from patches of the field, by erosion or by machinery, leaving exposed subsoil, these areas may be reclaimed by applying heavy doses of green mulch or manure, and then ploughing or digging this in. The ploughing will loosen up the subsoil and the organic matter will improve the structure. The field can then be farmed using minimum tillage, which will prevent recompaction, and large annual doses of manure or compost.

Drainage

Most crop plants cannot grow when the land is waterlogged for even part of the growing season. Swampy land can sometimes be salvaged by drainage, digging and regularly clearing ditches to remove standing water. These ditches should run directly to the lowest portion of the field. Where the field is very flat, the ditches can be gently sloped to prevent silting. Raised beds can be used to achieve the same effect as ditching.

CHAPTER 3

Weed Control

The immediate practical problem which faces the farmer is the control of weeds. This problem has to be tackled before the farm can be established. The advantages and disadvantages of six common methods of weed control are discussed below.

Hoe and cutlass

This traditional method is often used in conjunction with burning. The bush is first cutlassed down and left to dry for a short period before being burnt. At planting, the soil is loosened by hoeing, when any emergent weeds are cleared up. To farm well and to provide the crop with a weed-free environment is a daunting undertaking. One farmer can maintain about 0.3-0.4ha (0.75 acre) of land weed-free using a hoe and cutlass. The limitations of this scale mean that the farmer relying on the hoe for weed control cannot produce substantial amounts of food without massive amounts of labour.

Burning

This is often the only practical way the farmer has of clearing sufficient land to plant. It is quick and easy. The burnt residue of the vegetation provides some phosphate and small amounts of minerals, but all the organic matter and various nutrients are lost. Burning is obviously a one-off technique, it cannot be used to weed the young crop. Usually a hoe is used for this. Regular burning encourages the growth of resistant weeds, particularly *Imperata cylindrica* (Spear grass), which is difficult to get rid of after it has infested cultivated land.

11

Mechanical and animal-drawn tillage

Ploughing, harrowing and rotovating are all weed control measures, and if done well can be very effective. Tillage demands that the field should have been thoroughly cleared and all roots removed. It enables a much larger area to be made available for planting than is farmed by traditional farmers. On many farms where the tractor is hired or where a plough is the only available equipment, subsequent weedings have to be done either with chemicals or using manual labour, which will be very costly.

Ploughing can offer a method of controlling *Imperata cylindrica*. The Crops Research Institute (CRI) recommends two ploughings followed by a harrowing at the onset of severe drying conditions. This method has its dangers. If the ploughing is not done well and the plant roots are not cut into small pieces and left on the surface, or if dry conditions do not prevail, the weed may spread further.

The above three methods of weed control: hoeing, burning, and tillage, all remove weed growth from the soil surface. This is damaging to tropical soils, and greatly increases the risk of erosion and overheating to the soil. The aim should always be to reduce the time period over which the soil surface lies exposed to the bare minimum. It is better still to do away with bare soil altogether. The following methods have the important advantage that they do not expose the soil surface.

Chemicals

Chemical weedkillers are quick methods of weed control that can be used at the correct time and on young crops. Their use enables farmers to increase the area under cultivation. As chemicals are expensive both to the farmer and in terms of foreign exchange, they are only likely to be economical when they are used by a progressive farmer growing high yielding crops. The farmer must be prepared to use the weedkiller at the correct times to protect the young crop. Many different types of weedkiller are available. However, the minimum tillage farmer will be interested in leaving the soil with a mulch. This will be provided by weedkillers which kill growing tissue rather than those which act to prevent weeds from germinating.

Chemical use causes great concern nowadays. Many outdated pesticides, banned in their countries of manufacture are still available in Ghana. These are being used and abused at the moment. This book aims to inform of some of the dangers of pesticide use and to enable farmers to use recommended chemicals safely. In alley cropping the farmer is introduced to a method of farming that stabilizes the soil whilst reducing the need for pesticides.

Mulching

The mulch acts as a weed suppressant by blanketing off the soil. Any plant germinating under a heavy mulch will emerge into a dark, dank environment which will quickly kill it. Normally a few weeds regrowing from old established roots will emerge. These can be cutlassed off. Mulch can be used to protect a vulnerable young growing crop. When the farm is being established, weed growth will be abnormally heavy and a disproportionate effort may be required to control it in the first season. Gradually a mulch layer will build up that reduces the heavy initial need for weed control.

Over a period of several seasons, the farmer will notice changes in the kinds of weed present on a minimum tillage plot. Grasses, particularly, will tend to replace the original weed species. Grasses regrow fast and are relatively resistant to Paraquat (Grammoxone) sprays, which are commonly used in minimum tillage. In future Glyphosate (Round up) may become more available. As well as being safer, this is more effective than Paraquat. Currently, because of the prohibitive cost, it is rarely used except against *Imperata cylindrica*, which is resistant to other chemicals. However the problem of resistant weeds is not just limited to Paraquat. It is common wherever one particular weed control method is relied on too heavily.

Alley cropping

An advantage of alley cropping is that it uses three methods of weed control. A canopy cover shades out weed growth whenever land is not being cropped. This makes pre-planting weeding easy. When the crop has emerged the leafy canopy is cut to form the mulch. Weedkiller can be used to kill the few weeds that do emerge. Because alley cropping ensures the soil surface is never bare, it prevents erosion and overheating. At the same time, weed control is more effective and cheaper than using chemicals alone. Alley cropping need not necessarily use minimum tillage as the farming technique. Mechanized tillage is also being practised in alley cropping plots at the International Institute for Tropical Agriculture (IITA), in Nigeria.

Alley cropping offers a cost-effective method of controlling *Imperata cylindrica*. By allowing the canopy to grow until the shade effect is complete, the weed will quickly die out.

CHAPTER 4

Using Chemicals on the Farm

Chemicals are widely used on many farms. Nevertheless there is an increasing debate about their use. Chemicals offer a quick way to increase the harvest, but there are hidden costs. Safety is perhaps the most obvious one. Minimum tillage is a way of controlling the weeds whilst protecting the soil from erosion. To do this it relies heavily on chemicals. These are expensive to the farmer. They are also dangerous, not only to the farmer but also to those working in the area, and they can poison the soil. A healthy soil is the key to a fertile farm. Many people believe that the soil is not made healthy by the application of chemical poisons. Any farmer who chooses to use chemicals must be prepared to treat them with respect. This will cost money and time.

The following guidelines, if followed by the farmer, will help to keep the farm a safe place to work.

1) It is particularly important that small children are not allowed near chemicals.

2) Know the pests (insect or weeds) that need controlling. Advice on the best methods of control may come from agricultural teachers and extension workers. Only use the recommended pesticide, the wrong one may harm the crop.

3) Read the label carefully. The label will contain information on the types of pest controlled, when and how to use it, and what concentration to apply. It will also give specific safety instructions. Avoid buying a tin without such a label.

4) Keep chemicals dry, safe, and locked up.

5) Have an assistant on the farm whilst spraying.
6) Never smoke, drink, or eat whilst spraying.
7) Use protective clothing, a long sleeved shirt, long trousers, and rubber boots are essential. They must be bought with the sprayer. Anyone mixing chemicals should wear gloves and glasses also.
8) Take extra care when pouring out concentrated chemical, do not splash it and stand upwind out of the smell. The chemical should be added to a sprayer already half full of water (so that accidental splashes are diluted).
9) Keep soap and water handy to wash off any splashes immediately. Wash when the spraying is finished.
10) The chemical label may give a preharvest date, it may also give an access date. These are important. The preharvest date states the time interval that must pass after spraying before the crop is safe to harvest and sell or eat. The access date states when it is safe for people and animals to return to the sprayed area.

Spraying equipment

If chosen with care, a sprayer can be a useful tool in minimum tillage farming. The wrong choice can greatly limit the uses of the sprayer. For most small-scale farmers in the forest zone of Ghana, the tractor-mounted sprayer is not available. Some farmers can hire a tractor but its availability and serviceability tend to be unpredictable. For these reasons many small farmers use manual sprayers. There are three types commonly available.

The mistblower

This motorized sprayer is often seen being used by cocoa farmers. It produces a fine mist of spray which can reach a considerable distance. This type of sprayer is completely unsuitable for minimum tillage farmers as it has no fine control over where the spray will land. When using a weedkiller next to a field of standing crops this can cause their destruction. **Some weedkillers such as Paraquat (Grammoxone) produce a lethal spray mist and must never be used with this type of sprayer**.

The controlled drop applicator (CDA)

This machine produces a fine mist of spray by means of a rotating disc which throws droplets off itself. The mist will cover the plant effectively without using a great quantity of water. Again, because of the fineness of the mist these sprayers cannot be used with certain chemicals such as Paraquat. Under conditions of even a slight breeze, these machines are not accurate enough for work with weedkillers. The CDA machines have a future where water is scarce and where

Fig. 1. A mistblower　　　　　　*Fig. 2. A CDA sprayer*

knowledgeable advice is readily available. However, the lack of robustness of current machines makes them of little interest to most farmers.

The knapsack sprayer

For the minimum tillage farmer, this is the most versatile sprayer available. Usually of 15 to 20 litres capacity, it is designed to be carried on the operator's back, leaving the hands free to pump up

Fig. 3. The knapsack sprayer in action

17

pressure and guide the spray jet. This machine is ideal for minimum tillage farming as it can apply many different types of chemicals.

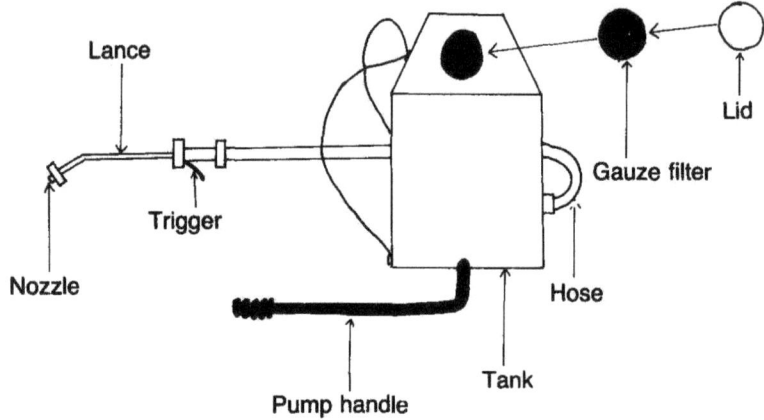

Fig. 4. Parts of a knapsack sprayer

The sprayer applies expensive chemicals, so it is in the interest of the farmer to ensure that the machine is not leaking or overapplying the chemical. A common problem with knapsack spraying is caused by the use of dirty water containing grit. If water is taken from a muddy stream or well and allowed to stand for a day before use, the grit will settle. No matter how clean the water, it is always essential to keep the gauze filter in place when putting water into the tank, failure to do this will shorten the life of the sprayer. Another common problem is blockage of the nozzle with grit. If the sprayer becomes hard to pump and the flow is reduced, this may be the cause. The nozzle assembly should be unscrewed and washed in soapy water. Never blow the nozzle clear with your mouth as it is poisonous. If the blockage persists, tease it out with a sliver of wood, metal should not be used as it will scratch and damage the nozzle irreparably. Sometimes, leaving the assembly to dry can help. When using a VLV nozzle (see below), always use the gauze filter supplied, or continual blockages will result.

Nozzle types

Many different types of nozzle are available, each for a different purpose. When applying weedkillers, a thorough covering of the vegetation with heavy droplets and little drift is the ideal. For this, the flat fan nozzle is recommended. A commonly available range of this nozzle comes in a series of different sizes, each size a different colour. Those nozzles with a large bore (hole size) will spray more packs, using a weaker mixture over a set area than those nozzles with a smaller bore. This is used where weeds are thick and dense, to ensure that all the vegetation is wetted. The nozzles with a narrower bore are more economical of chemical and water, and can be used on small weeds. The different nozzles also have different spray widths or swathes. Thus a nozzle with a wide swathe would be used to cover a large area of ground, whilst one with a narrow swathe would be used for small areas such as between the crop rows.

Nozzle Size	Packs/ha (acre)	Litres/ha (acre)	Width of spray (m)
YELLOW	15 (6)	227 (92)	0.5
GREEN	10 (4)	151 (61)	1.0
BLUE	12.5 (5)	185 (75)	1.5
RED	13 (5.5)	196 (79)	2.0

Table 1. Approximate output of polijet nozzles, all figures at a speed of 3.6 kph (1 metre/sec) and pressure of 1 bar. Swathe width with nozzle held at knee height.

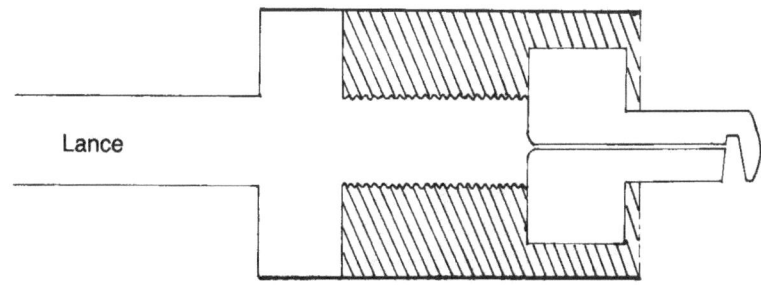

Lance

Fig. 5. Section through a flat fan nozzle

There is also a Very Low Volume (VLV) flat fan nozzle available. This has a very narrow bore and a wide swathe. It is used to cover large areas of low weed quickly but will only work on weeds 15 cm

(6in) or less in height. These nozzles are economical in the use of both water and chemical.

A commonly used nozzle for applying insecticides is called the swirl cone. It is designed to apply the insecticide thoroughly to all the plant leaves.

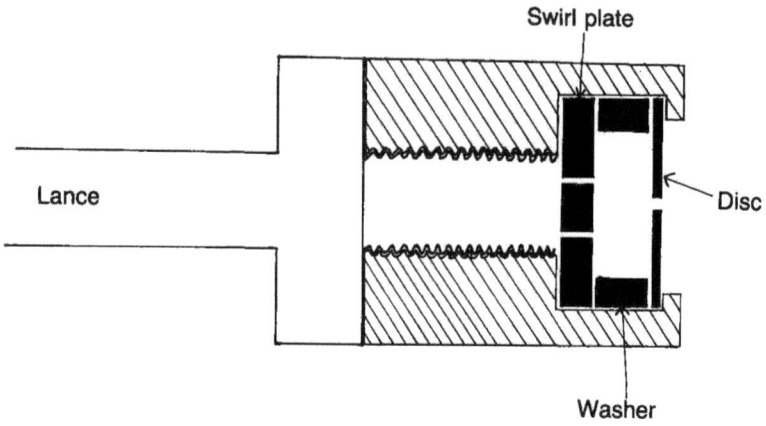

Fig. 6. Section through a swirl cone nozzle

The swirl plate is a disc with two or more slanting holes cut through it. On this rests the washer which creates the swirl chamber. The outermost visible portion of the nozzle is the orifice disc. These are stamped with a number to indicate the size of the bore. Disc size 12 is widely used.

Disc size	Output litres/hour	Litres/hectare at 60-70 cm row spacing
8	32-29	150
12	46-57	220
15	61-75	300
18	78-96	375

Table 2. Approximate outputs of different orifice discs used on a swirl cone nozzle. All outputs at 3 bar. Litres per hectare calculated on the average row spacing for cowpeas.

The spraying machine should be inspected regularly, and a small stock of spares such as valves and nozzles kept, to deal with problems that occur during spraying. When not in use, the machine should be kept in a dry, shaded place, secure from children.

Weedkillers

The choice of weedkillers can be bewildering. As they are expensive and only effective in the right conditions, it is necessary to choose the correct chemical. The following terms are sometimes used to describe weedkillers:

Selective: a weedkiller which kills only some types of plants. If used correctly it will kill weeds and leave the crop standing.

Non-selective: these weedkillers kill all the plants they touch.

Systemic: these chemicals kill all parts of the plant, including the parts underground.

Contact: these chemicals affect only the parts of the plant to which they are applied. Some plants may regrow from unaffected parts.

Residual: these chemicals affect plant growth over a long period.

Pre-planting: the chemical should be applied before planting the crop.

Pre-emergence: the chemical should be applied before the crop emerges.

Post-emergence: apply the chemical after the crop has sprouted.

Paraquat has been widely used in minimum tillage. It kills all green plants. Weeds can resprout, though crops rarely do! If used between the rows of a growing crop the plants must be protected by the use of a spray shield. This can be made from an empty paraquat container. Paraquat can only be used on weeds below knee height, if the weeds are bigger than this they should first be cut down with a cutlass.

Glyphosate is a non-selective weedkiller that kills all parts of the plant, so there is no regrowth. It is slow acting and expensive but it is safer to use than paraquat.

Atrazine is sold under many names. It is a selective weedkiller used with maize. It is absorbed by the roots of the plant. Whereas maize and a few other plants can break it down, most cannot and are poisoned. It has a residual effect of several weeks which can affect the growth of subsequent crops such as cowpeas. It should not be used on the same land more than once in a year.

Use of any weedkiller season after season and year after year is bad practice. Every weedkiller has weeds which are resistant to it, and the farmer will quickly find fields full of weeds which cannot be killed.

Insecticides

Insecticides are usually very poisonous to man. **In particular, the farmer should not use chemicals such as Aldrin, Endrin,**

Dieldrin and DDT, which are recognized to be highly unsafe.
Skilled farmers do not need to risk life and health by using the most dangerous types of chemicals they can find. Instead they will put the right chemical on at the right place, and at the right time, thereby killing the insects that are causing the trouble. The first rule is to find the insects that are damaging the crop, this is best done by walking through the field with someone who is knowledgeable about the crop.

Cypermethrin (Cymbush, Ripcord) is an insecticide which can be used against leaf feeding beetles and caterpillars on cowpeas and vegetables. It is recommended for use against Thrips and Pod Borers on the flowers and young pods of cowpeas.

Dimethoate (Roxion, Rogor) is a systemic insecticide taken up by the plant and effective against sap sucking insects such as aphids, pod sucking bugs, and the leaf hoppers which transmit Maize streak virus (see Chapter 7).

Practicalities of spraying

Some spraying machines have a pressure switch fitted. Insecticide spraying should be done under high pressure, weedkillers should be applied using a low pressure.

When calculating the correct concentration of chemical to add per

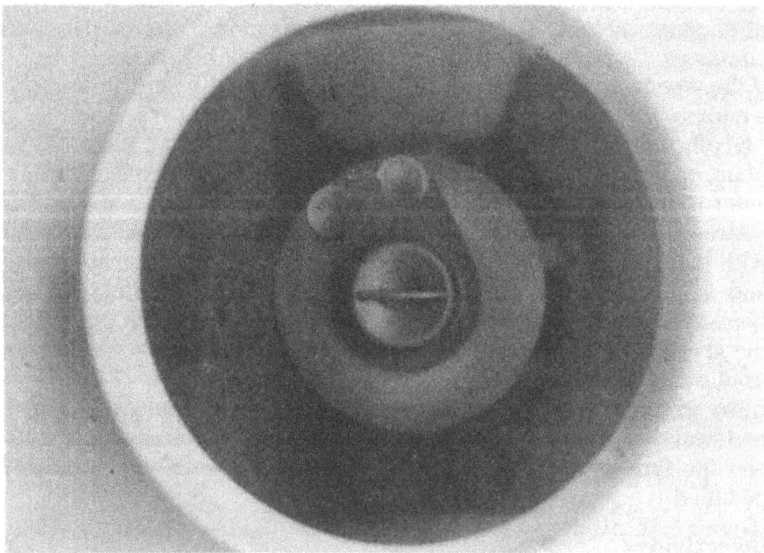

Fig. 7. Photograph of pressure switch as fitted to CP 15 knapsack sprayer

pack of spray refer to the label on the chemical container. Some sample calculations are given at the back of this book.

Application rates for certain chemicals commonly used in minimum tillage are given below. These rates are for a 15 litre sprayer, they are slightly lower than the manufacturers recommendations and are for general guidance.

Paraquat

Yellow polijet nozzle = 80-150ml/pack.
Green polijet nozzle = 120-150ml/pack.
Blue polijet nozzle = 100-120ml/pack.
Red polijet nozzle = 100-120ml/pack.
VLV 50 nozzle = 375ml/pack.
375ml is the highest concentration of paraquat that should be mixed into a 15 litre sprayer pack.

Roxion

Disc size 8 = 50ml/pack.
Disc size 12 = 35ml/pack.
Disc size 15 = 25ml/pack.

Water

Chemical

Water

Fig. 8. Sequence of diagrams depicting safe filling of the sprayer

23

Cymbush

Disc size 12 = 25ml/pack.

Ripcord

Disc size 12 = 75ml/pack.

Filling the sprayer

When filling the sprayer, add water to half full, add the chemical and then fill the remainder of the tank. A spray stool is a useful aid when putting on a full sprayer. This is a four legged stool, about three foot high, with its legs firmly embedded in the ground. The full sprayer pack should sit on it and slip easily on or off the back of the operator.

1) The sprayer is half filled with water.
2) The correct amount of chemical is added.
3) The remaining water is added.
4) The spray stool is used for loading.
5) Hold the lance so that the tip is horizontal.
6) Pump up the sprayer with four slow, steady, strokes to build up pressure.
7) Then press the trigger and start walking, an ideal speed is 3.6 kph (1 metre per second). This speed allows the operator to concentrate upon the work, pumping slowly and steadily while walking.

Storage

For stored produce, it is particularly important to apply only recommended storage chemicals, otherwise anyone eating the produce can be poisoned. Actellic is designed for use on stored maize and beans. It is sold as a dust which is easy to handle and does not require spraying equipment.

CHAPTER 5

The Practice of Minimum Tillage

Clearing the bush

Bringing a field into cultivation for the first time takes a lot of work. An early clearing of large brush from the field allows time for efficient weed control. It is in the farmer's interest to clear it as thoroughly as possible as, afterwards, the field may be cropped year after year with minimum effort.

The best time to clear is at the end of the wet season. Big trees should be cut and the wood disposed of, smaller bush can be slashed and left to rot down. Roots can be stumped, dug or burnt out. Chemical stump control is very expensive. As the dry season progresses there will be little new weed growth, any that does occur can easily be controlled at the start of the next season.

Weed control before planting

Work done at the TCC farm has shown that when weeds are controlled at the start of the major season, three separate weedings are necessary to give good control before planting. These weedings should take place approximately 4-5 weeks before planting, 2 weeks before planting and at planting. Some post-emergence weeding may be necessary. Weed control is generally easier after the first crop has been taken.

The first weeding

Some weeds will regrow during the dry season, though they will be small and weak. However, the coming of the first rains will change this, and the weeds will erupt. When they are 15-20cm high (6-10 inches), young, green and growing fast, this is the best time to control them. If using chemical control, use the red or blue polijet nozzle (wide bore) for the first weeding when the weeds are usually fairly dense, in subsequent weedings the green or the VLV 50 nozzle can be used.

The second weeding

After further rains more weeds will occur, particularly the fast growing weeds such as grasses and stump regrowths (which should be cutlassed off). As the weeds reappear and the field turns an even green, this is the ideal time to recontrol them. A second application of chemical or a light surface hoeing will be effective.

Fig. 1. Photograph of the mulch

At this stage the farmer will begin to notice the build up of a layer of dead weeds on the soil surface. This is the start of the mulch. The thicker it is the fewer weeds will be able to regrow through it. In places where a thick heavy mulch is built up before planting, as when cropping a field for the first time, it is necessary to cut a thin line through the mulch to plant along. These lines should be cut across the slope and at recommended row distances (see below). They can be cut after the second spraying has taken effect. Sighting posts can be used to ensure a straight line is obtained. The field is now ready for planting.

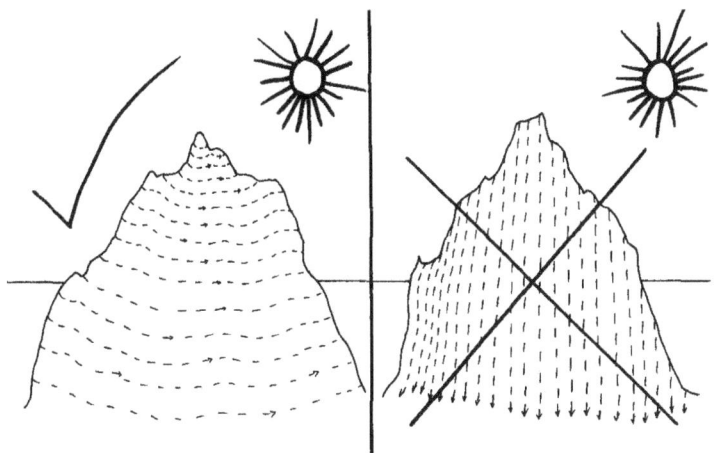

Fig. 2. Cutting lines along the contours

A third weeding may be necessary at or just after planting. If using Paraquat, the weeding must be pre-emergence of the crop. Weed control for subsequent plantings will be easier than for the initial one, as the previous mature crop will have shaded out many weeds. The crop residue can be used in place of the weed residue to form a mulch. Some farmers will choose to use a systemic chemical such as Glyphosate (Roundup), to control weeds pre-planting. These should be used 12 days before planting to give an effective kill. When using a residual type of weedkiller based on Atrazine, it should be applied at planting, or, if the label states, post emergence of the crop. Pre-planting weeding should be as described above. The Atrazine weedkillers can be applied with Paraquat, at planting, to provide protection for the young crop.

When to plant

The decision when to plant is a crucial one which, made correctly, can earn the farmer a healthy premium whilst other produce is scarce. The four points to look for are:
1) Harvest dates — are the conditions suitable at that time?
2) What is the current and likely future state of the rainfall?
3) Is the soil of the seed bed fit for planting?
4) Is the seed bed weed-free?
   The decision will be affected by the time taken by the crop to reach maturity in good conditions. Maize needs rain at tasselling, cowpeas require a dry period post flowering to mature in. Think about the probable market conditions existing at harvest time. It may be

27

possible to take a gamble on early rains being sufficient, thereby collecting a high price for the crop. Another option is to plant at the ideal time and thereby lessen the risk of not collecting a good harvest. In part this decision will depend upon the soil of the farm. A farm near water, or with a good soil capable of retaining soil moisture is at an advantage. The use of a heavy mulch is a great help in increasing water retention in any soil, and is useful in the conservation of the moisture of early rainfall. The expected rainfall in future weeks will play an important part in the decision. A meteorological station (eg at an airport), an agricultural college, or someone with long experience of the locality may be able to help with a forecast. The soil conditions should be good at planting, moist, but no standing pools of water. The final check that the farmer should make is that the seed bed is weed-free, to prevent weeds re-emerging to smother a young crop. The early stages are vital in any crop, so it is worthwhile to lay down a mulch to protect the young crop.

How to plant

Good planting will bury the seed at an ideal germination depth at the correct spacing all over the field. Burying the seed will protect it from partridges and rodents, but burying it too deep can slow down or even stop germination. The correct spacing for the crop is designed to allow each plant enough space in which to grow, and will make the field as a whole produce the maximum yield. The best way of getting the correct spacing is by planting in rows with set spaces between plants within each row, and between the rows.

Days to Mature	Crop Variety	Poor soil		Good soil	
		Within row (cm)	Between row (cm)	Within row (cm)	Between row (cm)
120	Maize La Posta	45-50	90	40-45	90
120	Dobidi	45-50	90	40-45	90
110	Texpano	45-50	90	40-45	90
90	Sofita	45-50	75	40-45	75
75-80	Cowpea	20	70	20	70
60	Cowpea	20	60	20	60

Table 3. Crop spacing for maximum yield on Good and Poor Soils.

Experience at TCC has shown that a machine called the Rolling Injection Planter (RIP), which is made in Ghana, makes the work of planting much quicker and easier. This machine can plant a variety of crops at different spacings. It enables larger areas to be planted than would be possible by hand. It is also convenient to use in awkward areas such as small irregular shaped fields.

Fig. 3. Photograph of the RIP

The machine is hand pushed, it will plant into a light mulch, but heavy mulches should have lines cut through, at the correct spacing. The beaks of the machine will inject the seed into the soil, and a heavy wheel will follow, rolling over and closing the hole. These planters will plant in most soil types, but soils which are very heavy or wet may block the beaks. When using on a clay or loam, it is worth waiting for a day following rain to allow the moisture to soak in. The machines are made at Kaddai engineering workshop, at Suame magazine, Kumasi. The same workshop also makes the interchangeable seed wheels that enable the machine to cope with different seed types, sizes, and seed spacing. A cutlass can be used instead of the planter. The work is slower and more tiring, but if no planter is available, or the soil is too sticky it may be the only solution. The small slit made when planting the seed is the only tillage necessary for minimum tillage.

Post-planting weeding

To ensure a good yield, post planting weeding is necessary, for any crop. In cowpeas it is especially important. During the first 3 weeks after emergence, the crop should be weeded twice, once at 10 days, and again 10-12 days after that. If the weed growth is heavy and the inter-row spaces are fully green, Paraquat may be applied using a spray shield. If a heavy mulch is used, often only one light weeding is needed. Maize will require weeding at 10 days to 2 weeks after emergence, again unless a heavy mulch is used. At, or around silking (when the cob produces fine hairs), a light weeding may be needed to control climbing weeds.

Fig. 4. Photograph of inter-row weeding

Fertilizing the crop

Crops must be provided with nutrients for growth. In part the soil will provide these, but today's high yielding varieties need so much that the farmer must also provide some. How much depends on the crop and the soil. At the moment in Ghana, the following chemical fertilizers are available: NPK (nitrogen, phosphorus, potassium) 20:20:0, ammonium sulphate and urea. Organic manures and composts can readily be obtained, these have the advantage of increasing the levels of organic matter in the soil, thereby making the soil healthier year after year.

For convenience, application times when using manure can be at the same times as a chemical would be used. However a pre-planting application would be as effective. Currently, on maize at the TCC farm, we apply 11 tonnes/ha of one-year-old chicken manure (4.5 t/acre), at one week post emergence. This is spread along the crop row from a sack. Chicken manure can contain up to 50kg nitrogen per 1.5 tonnes, plus varying amounts of phosphorus, potassium, and other nutrients. After 5 to 6 weeks, the maize benefits from a further 1 bag/ha of ammonium sulphate. Cowpeas on poor soils benefit from being started off with a small amount of manure; around 3 tonnes/ha is sufficient.

Chemical fertilizer needs to be more precisely applied than manure. It is expensive and, unlike manure, overdosing the crop does not benefit the soil. Virgin forest sites, or forest fallowed for more than 8 years, need no fertilizers. Chemical fertilizers should be placed around 3-4cm (2in) from the crop plant. The practice of burying the

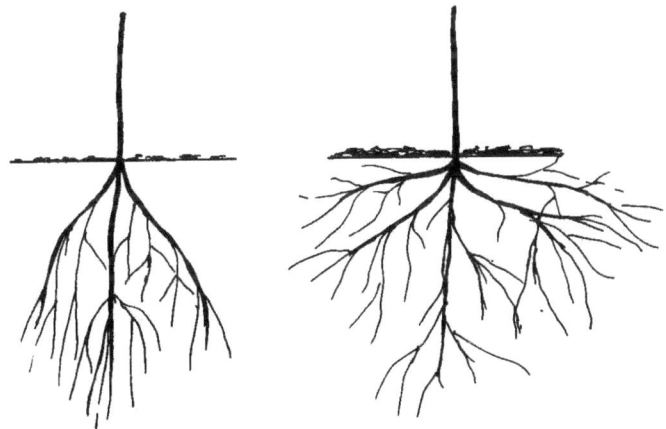

Fig. 5. Root growth with and without mulch

31

fertilizer is not necessary where a thick mulch is used. The mulch allows good root growth even at the soil surface, so a surface dressing is available to the plants. When harvesting the crop, leave as much residue as possible on the field to form the basis of next year's mulch.

Application	Compound	Old forest sites and Guinea Savannah	Alley Cropping and fallowed forest sites	Time of application
1st	20:20:0	5 bag/ha (2/acre)	2 bag/ha (1/acre)	Within 1 week of emergence
2nd	Ammonium Sulphate	2 bag/ha (1/acre)	2 bag/ha (1/acre)	5-6 weeks after emergence
3rd	Urea	1 bag/ha (½/acre)	1 bag/ha (½/acre)	not after 6 weeks.

Table 4. Fertilizer rates for maize in Ghana. Data from CRI recommendations, 1986.

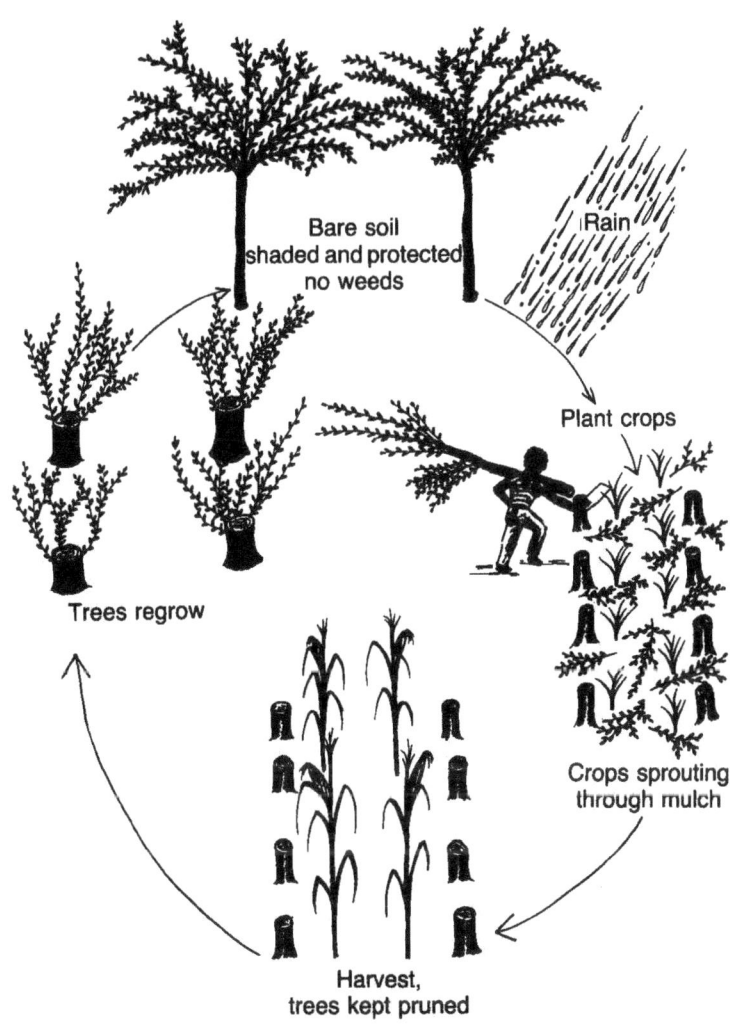

Bare soil shaded and protected no weeds

Rain

Plant crops

Trees regrow

Crops sprouting through mulch

Harvest, trees kept pruned

Fig. 6. The alley cropping season

33

CHAPTER 6

Practising Alley Cropping

Alley cropping is a special form of minimum tillage farming. It is particularly useful for the small scale farmer who is looking for a high yielding but stable form of agriculture, which is not heavily reliant on expensive agrochemicals. Alley cropping uses trees planted within the crops, to provide stability by maintaining soil fertility and thus giving a reliable high yield.

Fig. 1. Alley cropping at Approtech, canopy open

During the times of the year when there are no crops on the land the trees are allowed to grow until the leafy canopy is complete. A complete canopy will block out the light so that the field beneath the trees will have only a few unhealthy weeds. A seed bed may be prepared easily from such a field, either using weedkillers or a cutlass. The effort will be considerably less than when practising conventional minimum tillage. When the crop has started to emerge, the trees can be cut back as in Fig. 1, the leaves being used to provide mulch for the newly emerging crop.

Fig. 2. Alley cropping with a closed canopy

Most of the tree species used for alley cropping purposes are nitrogen fixing, and leguminous, this means that there are numerous bacteria associated with their roots. The bacteria cluster in small lumps of root tissue called nodules. The bacteria are able to absorb the nitrogen available in the air, and convert it to a plant available form, nitrates. Thus the tree is manufacturing its own nitrogen fertilizer. The nitrates are taken up by the tree and used to speed its growth. Correct management can result in other crops, grown in association with the trees, benefiting from the fertilizer. There are three main ways in which nitrates made by the trees can be made available to other crops. Clipping of the leaves and the rotting down of the resultant mulch will release the most significant amounts of nitrate. Under conditions of stress, the tree may shed some nodules intact. Such conditions might occur during a short dry spell in the growing season, or after a heavy pruning. There will also be a certain amount of leaching of nitrates from the trees nodules to the roots of other plants in close proximity.

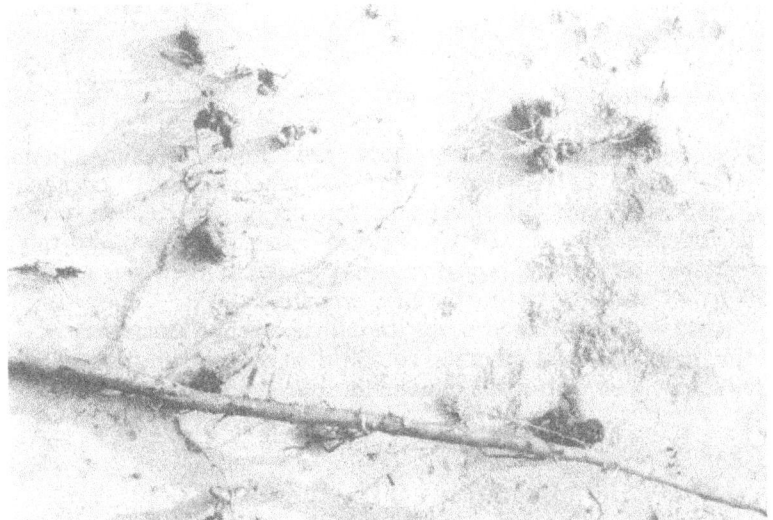

Fig. 3. Photograph of root nodules

Bacteria inhabiting the nodules belong to the genus Rhizobia. Normally a specific species of leguminous plant can only host a particular species of bacteria. If the plant is new to an area, as soya beans are to most of Ghana, then it is worthwhile inoculating the soil where the new crop is to be grown, with the correct bacteria species. This is true of many improved Leucaena tree varieties brought in from overseas. Inoculation may be done quite simply by taking a small bag of soil from an area where the crop is already well established (and where dug roots show the presence of nodules). This soil can then be mixed in with the seed for sowing. Because the bacteria are so small, many are contained in a small amount of soil. From areas which have grown a particular legume for many years there may be up to one billion bacteria per gram of soil. Healthy nodules can be distinguished by having pink, moist, interiors. They are attached to the root hairs rather than to the main roots, see Fig. 3 (cf root knot nematode galls, see Chapter 7). The trees will supply a considerable amount of nitrogen to the soil, but for the highest yields this should be supplemented with fertilizer, using the CRI recommendations for fallowed land, ie the reduced rate (see Table 4, p.32). Organic manure, if available, should be supplemented for the fertilizer.

Tree types suitable for alley cropping

Trees suitable for alley cropping share some or all of the following qualities:
1) Fast growing
2) Easy to raise
3) Deep rooted
4) Other special qualities, e.g. nitrogen fixing, or good fodder.

Two species of trees are commonly used for alley cropping in the humid tropics. These are *Gliricidia sepium*, and *Leucaena leucocephala*. Many other species of tree may be suited to the forest zone, or other areas. As alley cropping is currently being extensively researched, Appendix 2 contains some details of other tree species which may also be suitable for alley cropping.

Gliricidia and Leucaena both have advantageous characteristics, as the initial choice of planting stock will be growing in the field for a long time, it is worthwhile making the correct choice at the outset.

Fig. 4. Photograph of Gliricidia

Gliricidia sepium

The greatest advantage of Gliricidia is that it can be planted as a cutting. A one-metre stake cut from a mature tree, and put into the ground at the start of the rainy season will soon sprout and establish itself. Weeds cannot compete with this headstart. Gliricidia is a legume which grows fast and is a palatable fodder. It has a tradition of use in Ghana as a nursery shade for young cocoa plantations.

A disadvantage of Gliricidia is that it loses its leaves at the end of the dry season. For the first few rains there is little canopy cover, and this allows some weed growth. Nonetheless, the first rains seem to signify the time to put forth new leaves, so provided the branches are left on, the canopy is quickly restored, and the weeds cannot compete again. Unless planting early, this will not pose a great problem. The other disadvantage that Gliricidia grown from a cutting shows, and it is common to all plants grown from cuttings, is the lack of a tap root. A plant, including Gliricidia, which is grown from seed will exhibit a normal, deep main tap root. The advantages of this are shown in Fig. 5.

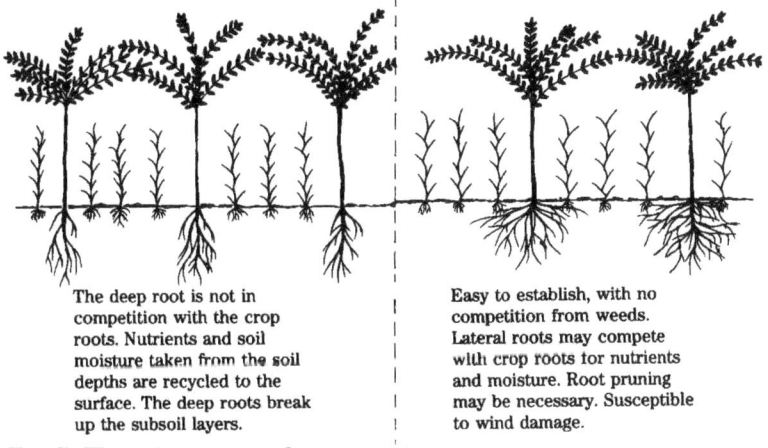

The deep root is not in competition with the crop roots. Nutrients and soil moisture taken from the soil depths are recycled to the surface. The deep roots break up the subsoil layers.

Easy to establish, with no competition from weeds. Lateral roots may compete with crop roots for nutrients and moisture. Root pruning may be necessary. Susceptible to wind damage.

Fig. 5. The advantages of a tap root

Gliricidia leaves are an excellent fodder source, without the toxic properties associated with Leucaena. However the wood is not a good fuel: it does not store well and it burns quickly.

Leucaena leucocephala

Leucaena is the most common tree used in tropical agroforestry. There are many good reasons for this. With some varieties growing up to 30 feet per year, it is one of the fastest growing trees in the world. It is also a prolific nitrogen fixer (see Fig. 3, root nodules on Leucaena). The root shape is ideal. Leucaena is planted as a seed, and will send down a strong tap root. The tree can stand a heavy pruning, and, in the forest zone of Ghana, it has a leafy canopy throughout the year. The major disadvantage with Leucaena is that the young seedling requires considerable care and attention until it reaches 60cm (2 feet). To minimize weeding, the nursing can be

Fig. 6. Photograph of Leucaena

done in beds, but this entails transplanting nursed stock. Transplanting is both time consuming and laborious, and unless great care is taken whilst uprooting the trees, damage to the tap root will occur. This will affect the root growth in the same way as being planted from a cutting. The tap root is deformed and deflected, and excessive lateral root growth will result. Direct seeding is easier both on the tree and the farmer, however it will entail some weeding to get the trees past the vulnerable young stage. An effective method is to seed the rows under a growing crop of corn. A trench should be dug, about 2-3cm deep, the seeds can be sown along it 1-2cm apart. This close seeding will require thinning out to the desired spacing later on, but it will ensure that some trees survive to be thinned. Any necessary weeding around the young trees should be done by hand. Whichever nursing method is chosen, the Leucaena seedling requires more care and attention than a Gliricidia cutting, yet the resulting tree may be a better performer. Once established, Leucaena is aggressive and it should not be planted if the farm is likely to be left unattended for more than one season, Gliricidia is less likely to become a weed problem in the same circumstances. Leucaena leaves have a limited use as a high protein fodder. Only a small proportion (3 per cent) should go into the diets of animals such as sheep or goats, because of problems with amino acid toxicity causing baldness. A similarly small proportion in the diets of layer birds will give the egg yolk a pleasing deep yellow colour.

Leucaena seeds, if harvested when the pods are mature, will require scarifying. The seeds need immersion in water just off the boil for 30 seconds. After this the water can be cooled by the addition of cold water. Failure to scarify will result in many seeds lying dormant after planting. Over-scarifying will result in the seeds being cooked, so it is important to keep the time to the minimum necessary.

Tree spacing

The number of trees on a field is a compromise between efficient weed control and competition for light, water, and nutrients (other than nitrates) between crops and trees. The ideal inter-row spacing for Gliricidia and Leucaena seems to be 3m, the within row spacing may be up to one metre. The TCC farm uses spacing of 3m by 75cm which allows the canopy to be established early and quickly. A larger between row distance is used on mechanized alley cropping farms.

Pruning regime and height

Growth of the trees should never be allowed to interfere with the crops' requirement for light. The important exception is during the early growth of the trees, so for the first year they should not be pruned at all. The crops may suffer but the trees must be allowed to establish. In the second season a light pruning may be needed. In subsequent years, the very early growth of a crop of cowpeas or maize will not be harmed by canopy shading, though this should be cut back within one week of crop emergence. Around Kumasi pruning appears to be necessary 4 to 6 times in each growing season. If the trees are grown for firewood or fodder, different cutting rates will be necessary. Generally, the more pruning is done, the more leafy growth will result and reduced cutting will produce more woody material. For mulch purposes, the tree should be cut at one metre (3 to 4 feet) high, as in Fig. 1, and the branches cut back to within 15cm (6 inches) of the main tree trunk.

When farming yams, the trees will save a lot of labour as *in situ* staking poles. They should be allowed to grow to the normal height for a yam stake before planting the yams. Just before planting, the trees are ring-barked at 15 to 20cm above the ground. As the yam begins to climb, the upper-parts of the tree will die and begin to shed leaf. The resulting stake is used by the yam for support, and the tree will slowly resprout from beneath the ring barking. This use for alley cropping has been tried and tested in Nigeria at the International Institute for Tropical Agriculture (IITA).

Alley cropping can be used to provide a cheap, reliable method of controlling problem weeds such as *Imperata cylindrica* (spear grass).

41

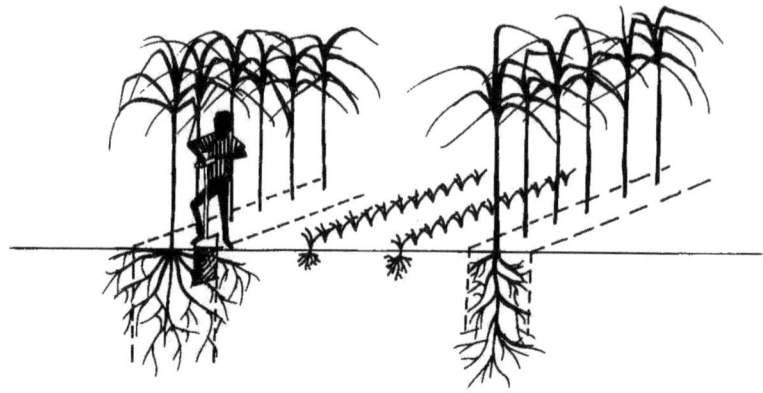

Fig. 7. Root pruning in progress

The weed cannot compete with a lengthy period of shade from the canopy of the trees. Once the Imperata has been controlled, it will not reinfest a well managed alley cropping plot.

It may prove necessary to prune the roots of the tree, if they are seen to be spreading into the cropped soil. This should be checked every 2 years. Pruning is a simple operation using a spade to cut the soil down to 20cm (9 inches) depth, at 25cm (12 inches) out from either side of the row of trees.

The alley cropping plot will require protection from bush fires throughout the dry season. The fire break should be a 3 to 4m (4 to 5yd) wide gap cut around the boundary of the alley cropping field. It should be weeded clear at the end of the rains, and maintained free of dead, dry plant material until next seasons rains. Only a severe bush fire will cross such a gap.

Agroforestry is a traditional form of agriculture in Ghana. Villages in the north have protected their *Acacia albida* trees, because they are legumes which enrich the soil below the tree, giving visible improvements to crop yields. Recently, alley cropping has been getting much more research attention, both from international centres such as IITA, and from research stations within Ghana. Its use in preventing erosion, as well as improving soils and yields from subsistence plots, and solving the need for fallow, has been demonstrated. It is now up to the farmer to take advantage of these benefits.

CHAPTER 7

Disease and Pest Management

This chapter describes some of the commoner pests and diseases of maize and cowpeas in Ghana. It is not an exhaustive list. Within Ghana more specific identification and guidance may be sought from the Crops Research Institute, CRI, at Kwadaso, Kumasi.

Maize

Maize has a number of pest and disease problems in Ghana. Generally their incidence is reduced in early main season crops, and more severe in late and minor season plantings. However each year is different and crops should be monitored for signs of trouble during early growth.

Maize stem borers

Three species are important in Ghana; *Sesamia botanephaga* and *S. calamistris*, *Busseola fusca*, and also *Eldana saccarina*, which tends to bore into the cob. The adults of all these species are moths. It is the larval stage, the caterpillar, which causes the damage.

Sesamia species. The eggs are laid by the moth in the leaf sheath. On hatching, the larvae will bore straight into the stem and then straight up to the top of the plant. The plant can collapse and growth may stop. The caterpillar is pinkish in colour.

Busseola fusca. The eggs of this species are laid in a similar position to those of Sesamia, on hatching the caterpillar climbs the outside of the plant to the growing whorl. It then eats its way down into the young leaves, finally killing all growth. These caterpillars are yellow in colour.

Eldana saccarina. Adult moths lay eggs on the silk of the developing cobs. On hatching, the young caterpillar will bore into the cobs, where it eats the grains and will increase the chance of secondary infections. The caterpillar can be distinguished by its grey colour.

All these species are important and destructive pests of maize. Their effect is most severe on second, or late first season maize. After the dry season, only a few individuals remain alive on alternative host plants, and the small amount of corn that is grown close to water through the dry season. The first maize of the new season provides a chance for these small populations to grow. Damage to subsequent crops grows with the number of adult stem borers laying eggs. Late and second season sowings can be protected from Sesamia and Busseola, by two applications of an insecticide at 6 to 8 days, and again at 3 weeks after emergence. Cymbush is suitable for this purpose.

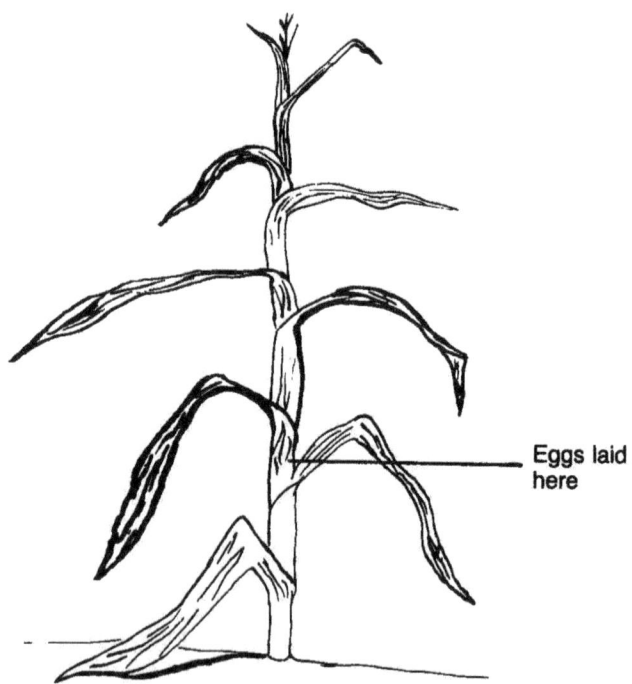

Eggs laid here

Fig. 1. Egg laying position on a maize plant

Fig. 2. A Streak damaged plant

Maize streak virus

This virus gives plants a green and white streaking along the leaves. If infected early the plant will be stunted and will not yield well. A late infection, at or after tasselling will have little effect on yields. Streak incidence is usually worst in second and late first season plantings. Monitoring the incidence in the first maize crops can enable the farmer to predict the likely incidence in later plantings. The virus is spread by a group of insects called leafhoppers. These pierce the plant with their mouthparts to suck the sap. The disease which they carry is injected into the young maize plant in a similar fashion to a mosquito infecting man with malaria. Control of the leafhoppers for the first 6 weeks of crop growth will tend to reduce the incidence of the virus, a systemic insecticide such as Roxion is effective against leafhoppers. Any plants seen to be infected during the early growth of the crop should be pulled out and burnt, this will also reduce the spread of the disease. Streak resistant varieties are available from Ghana Seed Company. Currently these do not yield well but they may be the only certain way of obtaining a harvest when Streak is very bad.

Rodents

Thryonomys swinderianus, Cane rat (Twi-*akrantia*). The cane, or cutting grass rat, is both well known and well liked by Ghanaians as a source of bush meat. It is also an important pest in fields of standing maize. The animal is nocturnal, being most active from 3 to 5am. It cuts down and chews up the young plants. Older ones are cut down and the cob is shredded. The favourite habitat of the rat is tall grassland, where it can be caught in wire snares, or with spring traps, or by hunting with dogs. Eradication of the long grass around the farm may help deter cane rat attacks.

Euxerus erythropus, western ground squirrel (Twi-*amokwea*). The squirrels are daylight feeders, they live in holes at night, coming out to feed in the grass and crop fields during the day. Damage usually occurs on ripe cobs on plants which have fallen to the ground. The squirrel can be caught using the locally made spring traps or by hunting with dogs. The meat is well liked.

Cricetomys gambianus, (Twi-*kouswea*). This rat is a nocturnal feeder. During the day it shelters in holes, often excavated from the base of termite mounds. These daylight resting homes are frequently changed. Damage is generally to dry, fallen cobs. The kouswea can be snared with wires, caught with spring traps, and when the hole is located they can be smoked out. Again the meat is prized as bush meat.

Because many rodents have very large territories which are constantly changing, rodenticides (rat poisons) are not effective unless they are applied on a massive scale, ie regionally. The chemical costs a lot of money, and the animal that is killed is usually quickly replaced by another. As eradication is not feasible on individual farms, it is cheaper and far more satisfying for the farmer to catch the bush meat and be able to eat it, rather than being left with a poisonous corpse to dispose of. The small rodents which cause so much damage to stored grain are similarly impossible to eradicate on a small scale. As they have no other value it is better to prevent them from attacking the grain to begin with. See rat and mouse guards for grain stores (Chapter 8).

If rodenticides are to be used against storage pests, or anywhere where children and domestic animals are likely to be, it is necessary to put them in secluded places where only the rodents can take them. The chemical can either be put in the animal's hole, or a shelter can be constructed from a piece of split bamboo with the internal segments knocked out of it. This should be placed upside down over the rodenticide, and pegged down at both ends. It will create a place where rats and mice will shelter.

Quelea quelea, weaver birds. These birds are an important pest to many crops, principally because of their numbers. They feed and live in colonies, their nests are woven from leafy material and hung on

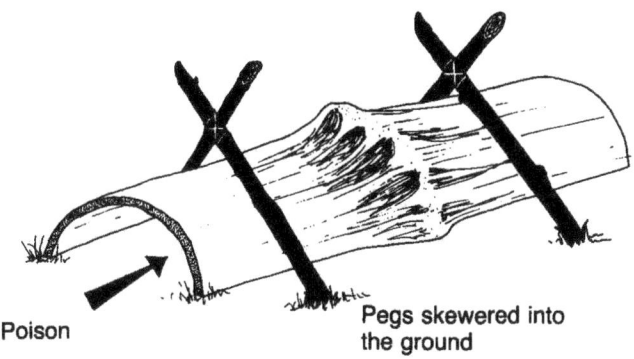

Poison   Pegs skewered into
              the ground

Fig. 3. Placing rat poison

tall grasses and other green plants. Much damage can be avoided by not planting the crop early. Alternatively, the birds can be scared by children. Nests should be cut down when found.

Sitophillus zea mais, grain weevil. The grain weevil attacks when the ear is mature. A closed ear is more resistant to attack than an open one. Many local maize varieties have tighter ears than do the improved ones. This has been recognized, and new varieties will soon be brought out with this defect remedied. The pest also makes use of damage caused by the weaver bird, and by *Eldana saccarina*, to gain entry to the ear. Once inside, the adult weevil drills holes in the grain, lays eggs, and multiplies fast. Because the initial infestation is in the field, it is good practice to dip the dehusked cob in 7.5 per cent Actellic solution, or alternatively to roll it in 2 per cent dust before putting it into store. Storage pests are dealt with in greater detail in the next chapter.

Cowpeas

Cowpeas suffer many insect pests, and control may be necessary several times after the start of flowering. Because of this, pre-flowering spraying should be avoided if possible.

At, and soon after emergence, cowpeas may suffer severe damage from a beetle genus, *Ootheca*. These insects eat the surface of the leaf, tracing a small path across the surface, which is later opened into a hole. The adult is reddish brown, sometimes darker, and about 8 to 10mm long. They are very common. It is only occasionally that they will actually threaten cowpea plants, usually when growth is delayed soon after emergence, as when a dry period affects it. In most years, the plants are growing fast and it is not economic to attempt to control the beetles. After 6 to 8 leaves have emerged, the plant is normally strong enough to cope with the *Ootheca*.

47

Fig. 4. Aphids on cowpeas

Aphis craccivora, aphids. Aphids can multiply very rapidly. They are commonly found on the stem and leaf stalks of the cowpeas. They occur as a black cluster, often tended by ants, if the cluster is brushed by the finger, small black insects will rub off. If the aphids are found to be infecting many plants before flowering, they should be controlled. A systematic insecticide such as Roxion is suitable for this purpose.

Flowering and post flowering pests

Megalurothrips sjostedji, thrips. The importance of this pest is not always realized, because it is small and the damage it causes is not always apparent until the insect has gone. It is a tiny black insect that inhabits flowers and buds. In very heavy infestations the insect can be seen swarming over the entire plant. The thrips can fly from plant to plant and are present on most, if not all, cowpea stands in Ghana. From early emergence of the flower buds the infestation will start. The insect feeds on the reproductive parts of the flower and the result can be a beanless pod. In heavy infestation, the flowers and leaf buds are distorted and will drop off. Management of this pest is achieved by a spraying regime of either Cymbush or Roxion, starting immediately the first flower buds are seen. This should be repeated every 12 days, or up to 4 days earlier if in the intervening period rains have been heavy.

Fig. 5. Thrips on cowpeas

Riptortes dentipes, pod sucking bug. There are several different species of pod sucking bug found attacking cowpeas in Ghana. *Riptortes dentipes* is the most common. A strongly built brown insect, about 20mm long, with well developed back legs and good flying ability, it can often be seen flying off when the crop is walked through. The mouthparts are long, and adapted for sucking the young green beans through the pod wall. The resulting damage is a pod containing shrivelled, worthless beans. The insect will be controlled by the 8 to 12 day spraying regime using Roxion that is applied to control the thrips.

Maruca testulalis (and others), pod borers. The larvae of these moths are very destructive and difficult to control. The adult moth is small and grey, and rests during the day on the underside of the cowpea leaves. The larvae of Maruca is whitish, with green sides, and brown dots. Another common species, Cydia, has a pinkish larvae. The eggs of Maruca are laid on leaf buds, flower buds, and in the flowers. The young caterpillar will bore in through the side of the flower, eating the reproductive parts, and leaving it dangling down by a characteristic thread of silk (see Fig. 7).

As the plants mature the caterpillar can bore into the stem of pods of the plant leaving a brown frass outside the entry hole. Inside the pod the caterpillar will eat the young beans. The caterpillar matures to the adult moth in the soil. Currently the best way to control this

Fig. 6. Riptortes dentipes

pest is a pyrethroid chemical. Cymbush works well and can be used when *Maruca sp.* is present.

In total there may be up to two sprays during flowering, and a possible further two during pod maturation. Where both Maruca and Riptortes are present it may be necessary to mix Cymbush with Roxion to achieve protection against both pests.

Grasshoppers

Zonocerus variegatus and others. At the end of the dry season this pest can appear in alarming numbers. It will feed on many green plants. As it is widespread little can be done, as any attempt at local control is swamped by numbers coming in from the bush. The population will decrease rapidly when the rains start regularly.

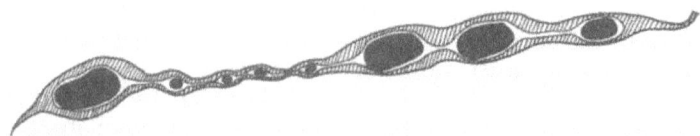

Fig. 7. Illustration of a damaged pod

Fig. 8. Maruca damage to a young cowpea flower

Nematodes

Nematodes are a problem especially where cowpeas are farmed intensively, or in succession with vegetables on the same land. Nematode damage is difficult to identify but it can be severe. The regular occurrence of one or more patches of stunted, or dying, cowpeas in a field may indicate the presence of a heavy nematode infestation. Nematodes are small microscopic worms living in the soil. Many species attack the roots of cowpeas. Where the damage takes the form of galls on the roots, these are often caused by the root knot nematode, *Meloidogyne sp.*, which is common in Ghana. The root forms galls around the nematode and infestation can weaken and kill the plant. Such galls can be distinguished from the normal, healthy, bacterial nodules found on cowpea roots by breaking them open. A healthy bacterial nodule is pink and moist inside, where nematode galls are hard root tissue. Also, bacterial nodules tend to be found attached to the tiny lateral roots, whereas galls are commonly part of the major roots.

There are many species of nematodes parasitic on cowpeas that do not form galls. Where plants look poorly, and when uplifted the roots are short, stunted, and prolific, nematodes may be suspected as the cause.

Management of Nematode problems. There is no effective way of

Fig. 9. Maruca damage to pod

curing heavily infested land, so preventive management should be the farmer's aim. Root knot nematodes can grow on a wide range of crops and weeds. The farmer should plan his cropping carefully so that populations of nematodes do not build up on susceptible crops in successive seasons. Such a plan is called a rotation. The ideal rotation should be: susceptible crop — poor host crop/resistant crop — susceptible crop.

Susceptible crops (crops on which nematode populations will increase) include tomatoes, garden eggs, okra, cucumbers and melons, all types of beans (including cowpeas), tobacco, bananas, pawpaw.

Poor host crops (crops on which root knot nematodes will remain alive but not breed) include cabbage, millet, sorghum, onion.

Resistant crops (crops on which root knot nematodes cannot feed) include groundnuts, maize, cassava and rice.

A good rotation might be, in the first season, maize or rice, followed by cowpea or vegetables, followed by groundnuts or maize or cassava or rice. Perennial plants, such as bananas and pawpaws, should not be grown in fields on which rotations are practised. They will support a high population of root knot nematodes to reinfect the crop from season to season.

When growing vegetables, it is good practice to ensure that the seedling at transplantation is free from infestation, as this will result

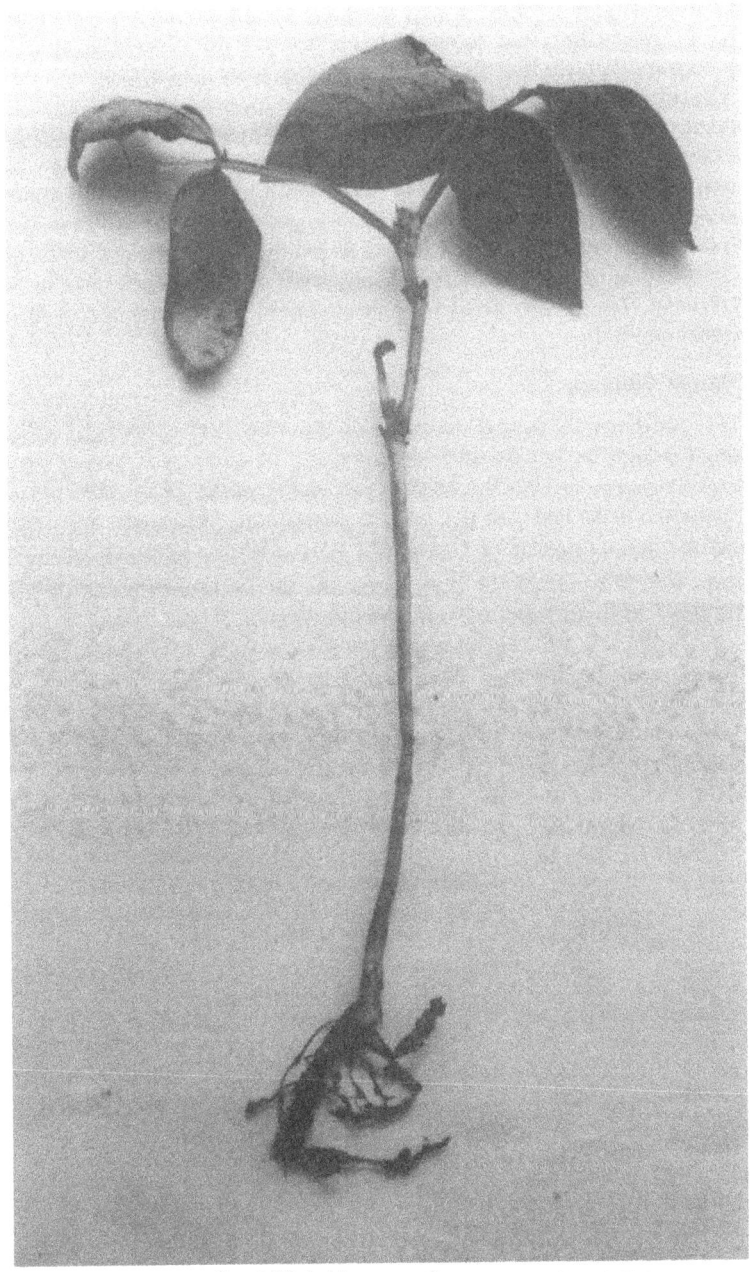

Fig. 10. Photograph of Meloidogyne damage on cowpeas

in a much higher yield. It can be done by either nursing the seeds on an area which has recently been flooded for several weeks, or by burning a large fire on the nursing area prior to seeding.

There is currently a nematocide on the market called Furadan. It is one of the most poisonous compounds known to man, and **its use without thick, protective rubber gloves should not be considered**. If the farmer is sufficiently knowledgeable about chemical use, and has protective gloves, and the wish to use Furadan, it should be incorporated (it is sold as granules) into the soil with the seed at planting. Some nematode-resistant crop varieties have been released. The Ghana Seed Company should be contacted to see if these can help.

Fungal disease

The incidence of fungal disease can be very high, especially when the weather is humid and warm. For cowpeas, the harvest plan should always aim for the beans to be harvested in a hot, dry, period. For maize it is vital that the crop is planted at a time when the rains will be good at tasselling. Unless the crop is to be sold for green corn, it is also advantageous that there should be a period of drying weather soon after the corn is harvested.

CHAPTER 8

Storage of Cowpeas and Maize

The fluctuations in price of cowpeas and maize are very considerable. A farmer can get a lot more for the crop if it is sold when it commands the highest price. For the year 1985, the farmer would have benefited most from selling cowpeas in March, and maize in June or July. To have cowpeas and maize available at these times would probably entail putting the 1984 harvest into store until the prices peaked. Stored crops have the disadvantage that the farm cash flow is held back, in the case of maize grown in 1984 and sold in 1985, for nearly one year. Storing crops is only possible when the grain is very dry. A good way to test the grain is to bite it. If it is very hard and breaks into smaller hard bits, it is likely to be dry enough. Failure to dry crops before putting them into store can result in heavy, sometimes total, loss due to fungal infection.

Cowpea storage

When drying in the field, care should be taken as some varieties will shatter on becoming dry, and the seed will be lost. The pods are best harvested, and then dried in the sun where fallen seed can be collected. In wet and humid weather, cowpeas are prone to heavy fungal attack or can germinate in the field if left too long. In wet weather the crop must be picked as soon as it becomes mature, to prevent field losses. It is for these reasons that minor season plantings are much safer for intensive cowpea farming. After harvesting, the pods should be dried in the sun, then threshed and the cowpeas can be dried again. This will help to control insect infestation before they go into store.

The main insect pest of stored cowpeas is the Bruchid beetle *Callosobruchus maculatus*, which can give heavy losses of up to 30 per cent in storage. The beetle first attacks the beans in the field. Eggs are laid on the surface of the beans, and look like small whitish caps. They hatch quickly and the larvae enter the seed, where they leave a characteristic hole.

In storage, many cowpeas are allowed to become heavily infested with *Callosobruchus sp.* beetles. The adult beetle chews small holes in many beans, and is very common. Since the infestation can start in the field, it is worth spraying the beans with Actellic before putting them into storage.

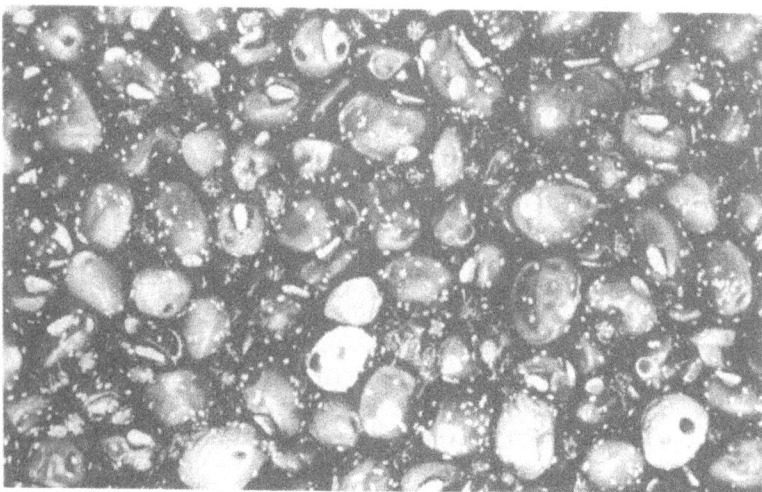

Fig. 1. Callosobruchus damage to stored cowpeas

The cowpeas can be protected by mixing 5ml palm oil per kg of beans (approx. one beer bottle per maxibag), which will give good protection. Another method is to mix ash with the beans, as burnt shell of groundnut provides an excellent insecticidal ash. It is full of sharp particles to scratch off the outer coat of the insect, which quickly dies. If using chemicals, 2 per cent Actellic dust is the easiest to apply: 0.5kg will treat 10 maxibags of cowpeas. A 7.5 per cent Actellic spray can also be used to spray the beans. **Other insecticides should not be used for storage as some, particularly DDT and associated compounds, will make the beans unsafe to eat.** Once treated, the beans can be put into storage. For the small farmer, a local, unglazed pot with a lid, such as are used for water cooling, is ideal. Once treated against field infestation, the pot will protect the beans for the remaining period of storage. Farmers with larger amounts of cowpeas may wish to use sacks, in which case retreatment may be necessary to prevent infestation. Whichever method is used, the beans should be checked monthly.

Maize storage

Maize needs to dry before it can be shelled and stored in sacks. The farmer growing small amounts of local varieties may leave the maize in the field for a long time after drying. Then it is tied in a bundle and put above the fire, both as a means of ensuring it is dry, and of using the smoke to drive away insects. If a farmer growing improved varieties tries leaving maize in the field to dry, it is likely to suffer heavy insect damage. Generally the new varieties do not have sufficient husk to deter insects. Often the grower of improved varieties is interested in replanting as soon as possible to maximize production, therefore he will wish to dry the maize off the field. If it is an early crop, and there is the possibility of catching peak market prices, grain drying ovens, if available, may be considered. For most of the harvest however, this will not pay. A method of drying under natural conditions is looked for. The FAO-type crib is designed to allow the farmer to pick the maize as soon as it is ripe, and minimize losses whilst drying for shelling.

When put into store, the cobs will need to be protected against insect pests. The most important pest is *Sitophillus zea mais*, the grain weevil. Characterized by a long curved rostrum or 'snout', this beetle is very common on grains throughout the tropics. Infestation starts in the field, where adults can fly from cob to cob. The adult female chews a hole in the grain, lays one egg per grain, and plugs it. The larvae develops within the grain and emerges in adult form leaving a characteristically damaged grain with a large hole in it. Adults are long lived, up to one year, and the female can lay many eggs.

For the farmer wishing to protect enough maize for the family, smoke from the fire may be the best protection. Farmers with more maize to protect may wish to use Actellic for this purpose, either the dust or the 7.5 per cent spray may be used but the cobs should be dehusked first. If using 7.5 per cent liquid, the most effective way is to put on gloves and dip each cob into the liquid. If using dust, either the cobs should be individually rolled in the dust or it should be sprinkled over the contents of the crib as each basket of fresh cobs is added. When in store, a fire lit beneath the crib will also help to protect the cobs, but any maize to be used for seed should be stored well away from the heat.

The FAO crib

This crib depends upon wind to dry the maize. Therefore the most critical dimension is the width through which the wind has to pass.

Fig. 2. The FAO crib

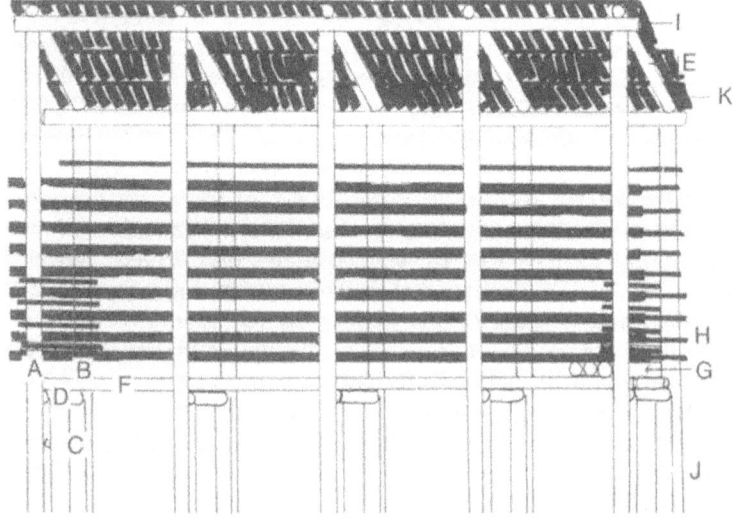

A = 5 × 4.7m   G = As necessary × 1.0m
B = 5 × 4.0m   H = As necessary × 4.5m slats
C = 10 × 1.5m   I = 2 × 4m pole.
D = 5 × 0.85m   J = Rat guards, 10 of.
E = 5 × 1.4m   K = Roof slats + covering.
F = 2 × 4m

Fig. 3. Construction of the FAO crib

The degree of rain protection provided by the crib is less important. The crib should be sited in a naturally windy area, and should be no more than 60cm wide (in the humid tropics) to allow wind passage through the cobs. In the semi humid tropics, one metre is the maximum internal width. If the crib is built wider than these guidelines indicate, the farmer is liable to suffer unacceptable losses during rains in August to October.

Materials for the crib. The crib can be constructed from any available wood. Bamboo, when used in Ghana's forest belt, will last 2 to 3 years. Teak can last indefinitely. For the roof, either thatching or metal sheeting may be used. Rat guards can be constructed from

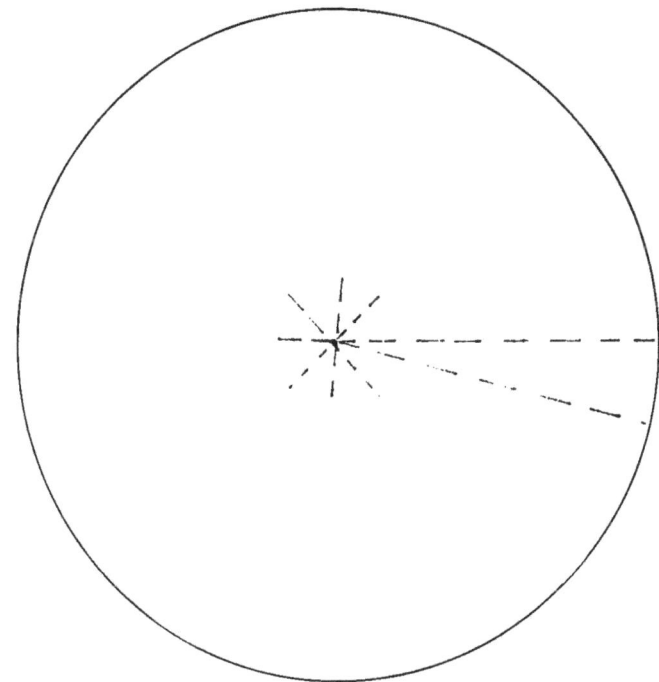

Fig. 4. Design for a rat guard

circular pieces of tin sheet, by cutting, as shown in Fig. 4, along the dotted lines. Fold around leg to make a cone.

Construction sequence A hammer and nails will be needed for the construction. All wood to go underground should be painted with a dirty engine oil/kerosene mixture. Holes should be marked out for the main frame as follows.

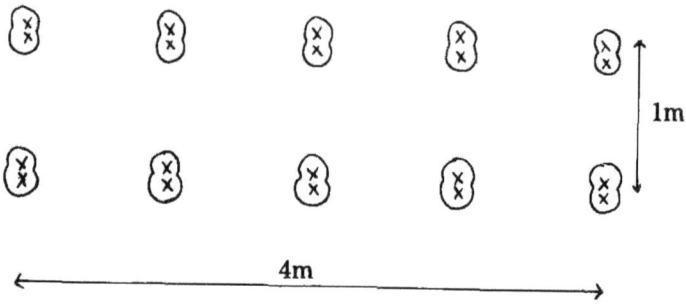

Fig. 5. Construction of FAO crib (1)

The holes should be dug to about 1m depth for the main frames, and 50cm depth for the basket support. When dug, the main frames (A and B), and basket support (C) should be erected vertically. See Fig. 6.

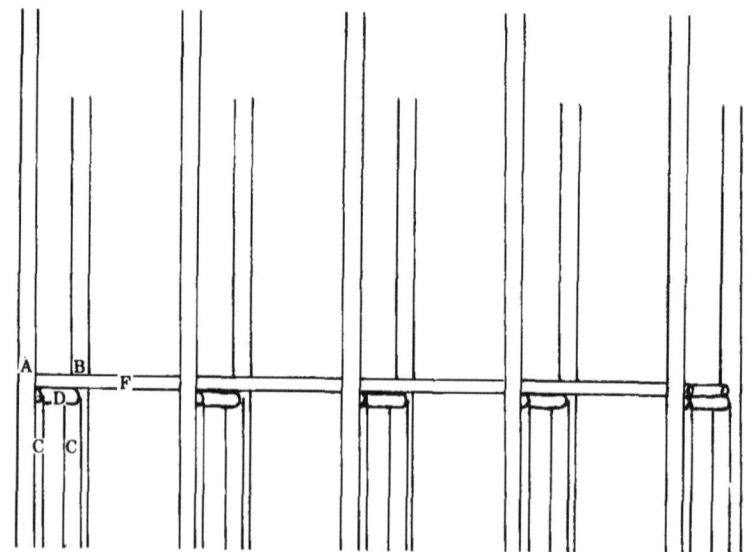

Fig. 6. Construction of FAO crib (2)

The horizontal basket supports (D) should be fitted to the vertical supports, and the two by four metre lenths (F) fitted adjacent to the main frame members. The two roof supports (I), followed by the sloping supports (E), can also be fitted.

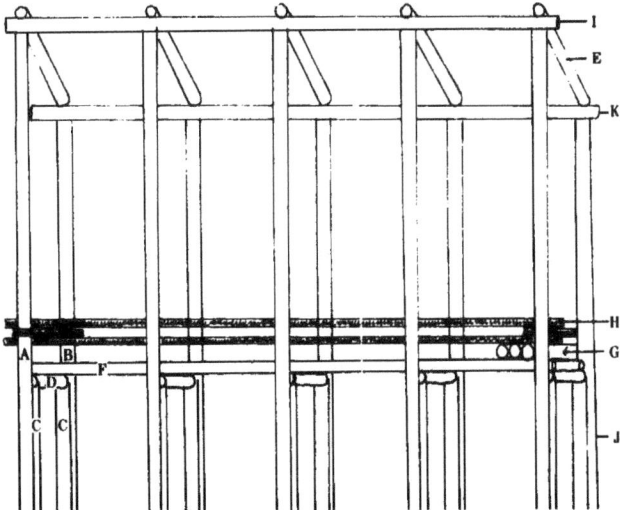

Fig. 7. Construction of FAO crib (3)

The slats (H) can now be nailed to the inside of the frame, leaving room for the floor poles (G). The best method of fixing the slats is to fit the two long sides, and then nail the end wall slats above those, the next two long-side slats rest on the end wall, until the basket is complete. Finally, the roofing and rat guards can be fitted. The rat guards should be 75cm off the ground. They are essential and it is necessary to weed the ground around the crib to prevent the rodents gaining access.

After November, the cobs can be shelled, when other farm work is slack. Shelled grain should again be re-treated against *Sitophillus* and the farmer can now effectively use ash, as with cowpeas. The sacks can be stored in the crib, with the walls covered to prevent any slight rains from entering. Before the next season's rains, the sacks of shelled grain should be taken from the crib and put under cover, since shelled sacks of maize are vulnerable to theft.

CHAPTER 9

Book Keeping, Cash Flow, and Raising a Loan

A farm, like any other business has to run to a financial plan. Following the plan entails good management. Estimates for the plan (cash flow) can be worked out with the help of someone conversant with current agricultural prices. It should allow for inflation and the constantly increasing price of goods and services. A contingency fund can be put aside to allow for unexpected costs, and to smooth out the effect of sudden price rises. In the plan outlay and income, each month over the year is detailed. Below, there is an example of a typical farm cash flow for a 4ha (10 acre) farm that is under intensive cultivation of 3 crops per year. During the construction of the plan, the general rule of taking upper estimates for outlay, and lower estimates for income was followed. This makes it less prone to overspending error.

The cash flow is invaluable in telling the farmer what scale of farming can be afforded. If a bank loan is needed, the first thing the bank will want to see is a cash flow plan. The bank may loan 'working capital' (up to the largest deficit amount shown in the cumulative cash flow), on the basis of the farmers 'fixed capital' (land and buildings). The fixed capital may be used to guarantee the loan in the event of the farmer not repaying. The cash flow will also tell the bank of the 'payback period'. This is the time the farm will take to produce a profit. The payback period is used to calculate the length of the loan.

Fig. 1. Cash Flow Table (all figures in Cedis) ◗

Planting time	Early cowpeas			Full season maize				Medium cowpeas						Early maize			Early cowpeas			Full season maize			
	Dec.	Jan.	Feb.	March	April	May	June	July	Aug.	Sept.	Oct.	Nov.	Dec.	Jan.	Feb.	March	April	May	June	July	Aug.	Sept.	Oct.
Cutlass 00	400												500										
Casual labour 70/day	5,250																						
RIP Planter 12,500	12,500																						
CPIS 4,000	20,000																						
Boots 1,500	6,000																						
Crib 600	600																						
Old drums 2,000	4,000																						
Grammoxone 300/1	42,000											39,000											
Roxion 500/1	3,000											3,000											
Actellic 400/1	400																						
Seed Maize (25/lb)			4,000			5,000								4,000						5,300			
Cowpeas (60/lb)	10,800								10,800							14,400							
Sacks 100	800		800							800			800								800		
Manure transport	1,000	500	500	500	1,000	500	500	500	500	500	500	500	500	1,000	1,000	1,000	500	500	1,000	1,000	1,000	1,000	1,000
Transport	7,500	7,500	7,500	7,500	7,500	7,500	7,500	7,500	7,500	7,500	7,500	7,500	7,500	10,000	10,000	10,000	10,000	10,000	10,000	10,000	10,000	10,000	10,000
Staff salary 2,500/mark	5,000	5,000	5,000	5,000	5,000	5,000	5,000	5,000	5,000	5,000	5,000	5,000	5,000	7,500	7,500	7,500	7,500	7,500	7,500	7,500	7,500	7,500	7,500
Manager salary 5,000/mark														5,000	5,000	5,000							
Contingency	2,500	2,500	2,500	2,500	2,500	2,500	2,500	2,500	2,500	2,500	2,500	2,500	2,500	2,500	2,500	2,500	2,500	2,500	2,500	2,500	2,500	2,500	2,500
Green Corn					50,240														100,000				
Cowpeas							3,200								112,460						65,879		

APPENDIX 1

Compost Making

For farmers without ready access to manure, compost offers a good alternative soil conditioning agent. It requires some effort to make well, but on small farms and gardens around towns it can be a profitable effort. Using the same materials, it is possible to make good or bad compost, the skill of the maker is an important ingredient.

Compost, or decomposed plant material, needs to rot down in contact with the air. For this reason, the size and height of the compost pit is limited, when making compost in the rainy season it should be stacked to a maximum height of 5ft. In the dry season the compost can be made in a pit up to 2ft deep to help moisture conservation.

Materials used to make compost

Green vegetation: needs no preparation.
Coarse dry vegetation: maize stalks, cobs etc, can be cut up into smaller pieces.
Fine dry material: straw, rice bran, etc. — no preparation required.
Sawdust: can be soaked in water for several days prior to use.

Urine, manure and wood ash all add greatly to the value of the compost and should be incorporated whenever possible. The compost heap (pit or stack) should have a layer of brush wood up to 5in thick at the base. This will allow aeration. The heap is then constructed in layers, for instance using a 10-inch layer there might be 3 inches of green vegetation, 3 inches of sawdust, pre-soaked, and 1 inch of manure with a sprinkling of wood ash and urine to finish. When the second layer is started the compost heap should have air vents incorporated into it. These should be positioned so that no part of the heap is more than 2 feet from air, i.e. they should be 2 feet from the outside edge, and a maximum of 4 feet from each other. The vents can be easily made by putting a vertical pole at the desired position of each vent, and as further layers are put onto the first layer the poles can be waggled to open a hole.

Some 10 to 14 days after the start of composting, the heap should be turned, edges into the middle. At this stage it can also be watered if it looks dry. A second turning after 3 weeks is desirable. At 9 weeks the compost should be checked to ensure that the rotting process has made all the materials unrecognizable. If this is not so it can be turned and watered again. By 12 weeks the compost should be finished and ready for application.

The above information is taken from; 'Composting in Tropical Agriculture', by H.W. Dalzell, IIBH, 1979.

APPENDIX 2

Calibrating a sprayer

Calibrating a sprayer tells the operator how much spray is being applied to the land, so the operator can work out the correct amount of chemical to put in the sprayer. To get a precise figure, it is necessary to mark out precisely an area of 10 metres by 10 metres. This requires a tape measure.

In an area of clear land put three sticks in a straight line, the first two must be 3 metres from each other, the third stick must be 10 metres from the first. Hold the tape in a loop at the 0 and the 12 metre mark so that the first stick is on the 0 and the 12 metre mark, and the second stick is at the 3 metre mark.

Use a third stick to hold the loop taut, and slide the stick around until it is at the 8 metre mark, put it in the ground. This stick should now be 4 metres from the first. These two sticks can now be used as a sight to place another 10 metres away in a straight line.

There are now two ten metre lines at right angles to each other. The same procedure can be used to mark out the remaining three corners. Once the 10 metres square is marked, use pegs to mark within it the operator's path. This will correspond to the swathe width for a flat fan nozzle and the row spacing for a swirl cone nozzle. The operator can follow these pegs to avoid spraying ground unnecessarily.

The operator will have to spray the land at a uniform speed, 60 metres per minute is recommended. This can be practised by walking at an even speed along a sixty metre line. It should take one minute.

1) Check that there are no leaks in the sprayer.
2) Hold the sprayer level and fill it with 10 litres of water. Do not use chemical for this exercise.
3) Put the sprayer on and stand at the start of the plot rows. Pump

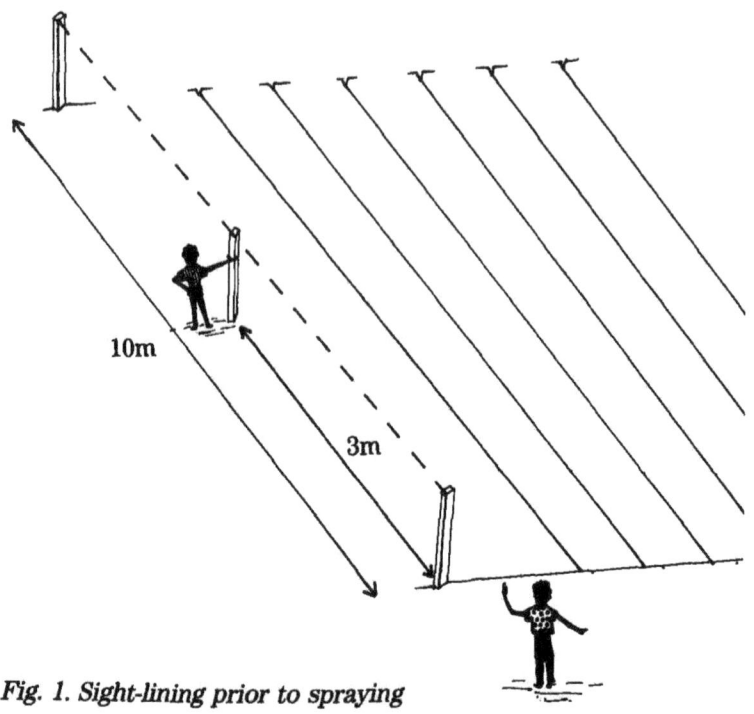

Fig. 1. Sight-lining prior to spraying

up the correct pressure and start spraying at a normal speed and
pumping rate. Only spray inside the plot.

4) When the plot is finished replace the machine on level ground
   and record the amount of water used.

5) The area is 1/100th of a hectare, so to find the amount of chemical
   that would be applied to a hectare, simply multiply the amount
   of water used to spray the square by 100. For example, suppose
   after spraying the square the machine contained 7 litres of water,
   therefore 3 litres have been applied: 3 x 100 = 300. This
   represents the flow rate, 300l/hectare, of the nozzle when used
   by that operator.

If the operator is aiming to apply a particular amount of spray per
hectare, different nozzles will give a different application rate.

Calculating tank concentrations

When calculating the correct concentration of chemical to add per
pack of spray, refer to the label on the chemical container. For ease of
calculation always use metric units, ie litres, hectares, and metres per
hour.

Below are some sample calculations:

1) An insecticide label states that, for aphids/thrips on cowpeas, 300 to 600ml of chemical should be applied per hectare. The field has some thrips and some aphids on cowpeas which are becoming leafy. A relatively high concentration of 450ml per hectare is chosen to give adequate coverage. A swirl cone nozzle, disc size 12, is being used. This gives approximate output of 220l/hectare, at a walking speed of 60 metres per minute. The sprayer tank capacity is 15 litres. The formula is:

$$\frac{\text{Chemical/hectare} \times \text{Spray tank capacity}}{\text{Nozzle flow rate}} = \text{Chemical required per tank.}$$

$$\frac{0.5 \times 15}{220} = 0.0351, \text{ or } 35\text{ml per pack.}$$

2) The label of an insecticide for stored products states that 200ml (0.2l) of 7.5 per cent solution should be applied to the contents of a 4 metre crib of stored maize: 200ml is too small a quantity for a knapsack sprayer, so a flit gun is employed. The insecticide is a 25 per cent solution. The formula is:

$$\frac{\text{\% concentration required}}{\text{\% concentration of insecticide}} \times \begin{array}{c}\text{total} \\ \text{amount} \\ \text{required.}\end{array} = \begin{array}{c}\text{amount} \\ \text{of chemical} \\ \text{required.}\end{array}$$